High Field Superconducting Magnets

High Field Superconducting Magnets

Fred M. Ašner

former CERN scientist

European Organization for Nuclear Research
Geneva, Switzerland

CLARENDON PRESS · OXFORD
1999

OXFORD
UNIVERSITY PRESS

Great Clarendon Street, Oxford OX2 6DP

Oxford University Press is a department of the University of Oxford.
It furthers the University's objective of excellence in research, scholarship,
and education by publishing worldwide in

Oxford New York

Athens Auckland Bangkok Bogotá Buenos Aires Calcutta
Cape Town Chennai Dar es Salaam Delhi Florence Hong Kong Istanbul
Karachi Kuala Lumpur Madrid Melbourne Mexico City Mumbai
Nairobi Paris São Paulo Singapore Taipei Tokyo Toronto Warsaw

with associated companies in Berlin Ibadan

Oxford is a registered trade mark of Oxford University Press
in the UK and in certain other countries

Published in the United States
by Oxford University Press Inc., New York

© Fred M. Ašner, 1999

First published 1999

A catalogue record for this book is available from the British Library

Library of Congress Cataloging in Publication Data
(Data available)

ISBN 0 19 851764 5 (hbk.)

Typeset using Latex
Printed in Great Britain by
Biddles Ltd, Guildford and King's Lynn

This book is dedicated to my wife Jagoda
and children Sandra and Igor.

PREFACE

The decision to write this book follows my 27 years of activity in the domain of super-conducting magnets, predominantly at CERN but also at the DESY, SSC—and LBL—laboratories. During that period I witnessed the continuous progress of the state of the art in superconducting magnets extending from the first 5 T quadrupole, probably the last specimen wound with an untwisted conductor, to the complex LHC-dipole proto-types, which safely attained fields up to 9.5 T.

In fact, after the fundamental work of the Rutherford Laboratory Group which trig-gered the development of fine filamentary NbTi and Nb_3Sn superconductors in the 1970s and early 1980s their widespread application has been euphorically forecast. Classical devices, however, so far remained for economic reasons and nowadays one hardly mentions superconducting transformers and generators or electrical energy trans-port by superconducting cables or overhead lines. Technical superconductors, however, found increasing application in domains where high magnetic fields and current densi-ties are essential: in superconducting accelerator and detector magnets and in NMR and NMI (nuclear magnetic imaging) devices for medical purposes. Future nuclear fusion and fission based energy generators will also belong to this group. These devices will exhibit high field precision, safe mechanical design and absence of premature quenching or training. To achieve this the active part of the magnet, its cooling and protection must be harmonized and their behaviour at different temperatures and stress levels matched and optimized. Putting it bluntly, a high field superconducting magnet of known EM-forces and an extremely strong outside clamping structure will fail and exhibit training if continual and efficient contact between all structural elements has not been assured. The present book has been written in this spirit.

To conclude, it is a pleasure to thank colleagues from a number of research institutes, laboratories and universities for stimulating discussions, reports, data and photos they kindly gave me. I especially acknowledge discussions with my former CERN colleagues Drs D. Leroy, J. Vlogaert, D. Hagedorn and S. Russenschuck; I am indebted to Dr S. Wolff from DESY, Dr C Meuris and Mr H. Desportes from CEA, Saclay, Prof. H. Ten Kate from the Twente University, Prof. L. Rossi from INFN, Milan, Drs R. Scanlan and R. Meuser from LBL, Berkeley, Prof. P. McIntyre from the TAC, University of Texas, Dr D. Markiewicz from NHMFL, Florida and to Prof. R. Fluckiger from the Geneva University for their valuable documentation. Last but not least I should like to express my gratitude to Dr M. Clayton of CERN who helped me with great patience to become familiar with the art of computer text and graphics editing, which I had to learn after retiring from CERN. Thanks also to Dr F. Dothan for his advice on acquiring the excellent Kaleidagraph program.

Genolier   F.A.
1998

CONTENTS

The plates fall between pages 172 and 173

NOMENCLATURE

Symbol	Meaning	Chapter
A, A_s	vector potential	2.3., 9.2.
A_0, A	initial and final wire cross-sections	5.2.
A_{Cu}, A_{tot}	Cu and total conductor cross-sections	7.2.
$A(s)$	maximum beam divergence	8.2.
a	composite radius	3.2.2.
a_f	superccconducting slab half width	2.3.1., 3.2.1.
a_n	skew multipole coefficient	9.2.
Δa_n	winding perturbation multipole coefficient	9.5.
α	Stekly parameter	6.2.4.
α_m	momentum compaction factor	8.2.
B_{c0}	critical magnetic induction	2.1.
B_{c1}, B_{c2}	lower and upper critical magnetic inductions	5.2., 5.6., 12.6.
B^*	conjugate complex magnetic induction	9.5., 13.2.1.
B_e	external magnetic induction	3.2.2., 5.2.
B_x, B_y, B_z	magnetic induction components in Cartesian coordinates	9.2., 12.2., 12.3.
B_ρ, B_θ	magnetic induction components in cylindrical coordinate system	9.2.,12.3.
B_0, B_{Fe}	bore and iron shield magnetic inductions	10.2.
B_{ref}	reference induction	13.2.1.
B_t	transport current induction	4.5.
B_{z0}	induction at current loop centre	12.2.
b_n	normal multipole coefficient	9.2.
Δb_n	winding perturbation multipole coefficient	9.5.
β	(adiabatic) stability parameter	2.3.1.
β_t	self-field stability parameter	3.2.2.
$\beta(s)$	phase function	8.2.
C	specific heat	2.2., 6.1.
C_e	electron contribution to specific heat	2.2.
$C(s)$	cos-like function	8.2.
Δc_n	winding perturbation multipole coefficient	9.5.
D_m	magnetic diffusivity	3.2.2.
D_{th}	thermal diffusivity	3.2.2., 6.4.3.
$D(s)$	dispersion function	8.2.
d_{eff}	effective filament diameter	4.5.
d_f	filament diameter	3.2.3., 4.5.

Symbol	Meaning	Chapter
d_h	hydraulic diameter	6.2.1.
δ	skin depth	2.3.
δ	differential thermal contraction	10.3.
E	electric field	2.1., 2.3., 8.1.
E	elasticity modulus	10.3., 12.3.,
E_0	transient input heat pulse	6.2.5.
$E(s)$	particle beam envelope	8.2.
e	electron charge	2.1.
ε	phase-space ellipse area,	8.2.
F	(composite) free energy	2.2.
F, Φ	complex potential	9.5., 13.2.1.
F_h	horizontal force	10.5.
F_m	mating face force	10.2.
F_{nc}	normal condition state free energy	2.2.
F_{sc}	superconducting state free energy	2.2.
F_s	longitudinal end force	10.2.
F_x	transverse Lorentz force	8.2.
F_z	axial force in solenoids	12.3.
F_ρ, F_θ	radial and tangential force components,	10.2., 12.3.
f	(unit volume) force	2.4.
f	quadrupole lens strength	8.3.
f_{pc}	(unit volume) pinning force	2.4.
f_b	EM body force	12.3.
f_p	body force due to diffential radial pressure	12.3.
f_s	repulsive force	2.4.
f_ρ	radial component of tensile force	12.3.
Φ	(quantized) magnetic flux	2.4.
Φ	complex potential	9.5., 13.2.1.
Φ_0	elementary fluxoid	2.4.
$\Phi(s)$	phase advance angle	8.2.
Φ_1, Φ_2	angles of sector winding	9.2.
G	heat per unit surface	6.2.4.
G_0, g	quadrupole field gradient	8.3., 13.2.1.
g	gap width	10.3.
γ	particle gyrometric ratio	12.4.
H	magnetic field	2.1.
H	Hall coefficient	13.2.2.
H_c	critical magnetic field	2.1.
H_{c0}	transition magnetic field	2.1.
H_e	external magnetic field	3.2.

Symbol	Meaning	Chapter
H_i	average magnetic field within the superconductor	4.2.
H_t	transport current field	2.4.
He I, He II		6.1.
h	Planck constant	2.4, 12.4.
h	heat transfer coefficient	3.2.2., 6.1.
h_f	film boiling heat transfer coefficient	6.2.
h_k^{-1}	Kapitza resistance	6.2.3.
h_t	transient heat transfer coefficient	6.2.
I	current	7.2.
$\pm I_c$	superconducting screening currents	3.2.
I_l	(superconducting) line current	2.1.
I_m	maximum recovery current	6.2.4.
I_{r0}	cold end recovery current	6.2.4.
I_t	transport current	3.2., 2.4.
I_1, I_2, I_3	interstrand currents in superconducting cables	4.4.
j	current density	2.1.
j_{av}	average current density	2.1., 10.5., 12.5.
j_c	critical current density	3.2., 4.2., 5.5., 7.1. 9.4., 10.2.
$j_{c\,non\,Cu}$	non copper crit. current density	5.5., 5.5.2., 5.5.3. 5.5.5., 5.6.
$j_{c\,av}$	average critical current density	5.5., 5.5.3., 5.6.
j_{cz}	axial critical current density	3.2.3.
j_s	longitudinal current density	2.4.
j_{sc}	critical current density in superconductor	2.3., 5.6.
j'	imaged current density	9.2.
K	heat conductivity	6.1.
K_x, K_z	accelerator magnet matrix parameters	8.3.
K_{tr}	transverse heat conductivity	7.2.
k	matrix thermal conductivity	3.2.2.
k	multipole strength	8.2.
ξ	radius of normal conducting hole	2.4.
ξ	normalized cooling parameter	3.2.2.
L	inductance (due to skin effect)	2.3.
L	heat of vaporization for He I	6.2.3.
L	period length of an accelerator structure	8.2.
L	angular momentum	12.4.
L_{eff}	effective coil length	13.2.1.
L_n	inductance of normal winding part	7.2.
L_{sc}	inductance of superconducting winding part	7.2.

Symbol	Meaning	Chapter
L_t	transposition pitch	3.2.3.
l_c	critical (twistpitch) length	3.2.3.
l_m	maximum uncooled conductor length	6.2.4.
l_{nm}	maximum normal zone length	6.2.5.
l_0	characteristic length	6.2.5.
λ	penetration depth	2.3.
λ	volume part of superconductor	3.2.2.
λ_{Cu}	volume copper part in superconducting wire	4.3.
$\frac{\Delta l}{l}$	thermal contraction	10.3.
M	magnetization	4.2.
M	transfer matrix	8.2.
M_θ	bending moment	10.2.
M_r	equivalent magnetization due to eddy currents	4.3.
M_0	magnetization in superconductor	4.2.
M_{slab}, M_{circ}	magnetization in rectangular and round superconductor	9.4.
M_{sc-n}	mutual inductance between superconductor and normal wdg parts	7.2.
m_e	electron mass	2.1.
m_s	Cooper electron pair mass	2.3.
m_0, m_i	mutual inductances	4.5.
$\dot{m}$	mass flow	6.1.
μ	magnetic moment	12.4.
μ	phase advance angle per cell	8.2.
μ_H	Hall mobility coefficient	13.2.2.
μ_0	vacuum permeability	2.1., 9.3.
μ_r	relative permeability	9.2., 9.3.
Nb_3Sn	niobium-tin	5.1.
NbTi	niobium titanium	5.1.
Nu	Nusselt number	6.2.
N_t	number of turns	13.2.1.
n	field variation index	8.2.
n	transition quality factor	5.2.
n_0	electron volume density	2.1.
υ	viscosity	6.2.2.
υ	Poisson coefficient	10.3., 12.3.
υ_L	Larmor or precession frequency	12.4.
ω	angular frequency	2.3.
ω_0	cyclotron frequency	8.2.

Symbol	Meaning	Chapter
P	cooling perimeter	6.2.4.
P	magnetic pressure	12.3.
Pr	Prantl number	6.2.
p	pressure	2.2.
p	particle momentum	8.2.
p	number of pole pairs	9.2.
p_m	mating face pressure	10.3.
p_0	central particle momentum	8.2.
p_ρ, p_θ	radial and azimuthal pressure components	10.2.
p_{tr}	transverse pressure	5.5.6.
Δp	pressure drop	6.1.
Q	heat flux	2.2., 6.1., 6.2.
Q	number of betatron oscillations	8.2.
$Q(T)$	cooling power per unit surface	6.2.4.
Q_b	burnout heat flux	6.2.2.
Q_{cr}	critical heat flux	6.2., 6.4.4.
Q_t	transient heat flux	6.2.1.
q	cooling per unit volume	6.3.
q_s	Cooper electron pair charge	2.3.
R	dissipation resistance	2.3.
Re	Reynolds number	6.2.3.
R_m	radius of measuring coil	13.2.1.
R_n	resistance of normal winding part	7.2.
R_{ref}	reference radius	13.2.1.
R_s	radius of concentric iron screen	9.2.
R_1, R_2	sector winding radii	9.2.
R_1', R_2'	imaged sector winding radii	9.2.
r_f	filament radius	3.2.2.
ρ	(metal) resistivity	2.1., 6.1.
ρ	radius of curvature	8.2.
ρ	winding radius	9.2.
ρ_d	dynamic component of resistivity	2.1.
ρ_e, ρ_{eq}	equivalent resistivity	4.3.
ρ_m	matrix resistivity	4.3.
ρ_n	resistivity of normal conducting part	2.3.
ρ_0	static component of resistivity	2.1.
ρ_n, ρ_s	normal and superfluid densities of He II	6.4.2.
S	entropy	2.2.
S_{eff}	effective coil surface	13.2.1.

Symbol	Meaning	Chapter
$S(s)$	sin-like function	8.2.
s	surface	2.4.
s	area between intersecting ellipses	9.2.
σ	electric conductivity	13.2.2.
σ_s	longitudinal or axial stress	10.2.
σ_θ	azimuthal bending stress	10.2.
$\sigma_{\theta m}$	maximum bending moment stress	10.2.
$\Delta\sigma$	stress overcompensation	10.2.
σ_{cr}	critical stress	5.5.6.
σ_1	longitudinal stress	5.5.6.
σ_ρ	radial stress in solenoids	12.3.
σ_t	tangential stress in solenoids	12.3.
σ_z	axial stress in solenoids	12.3.
T	temperature	2.1.
T_{cal}	calibration temperature	13.2.2.
T_{c0}, T_c	critical temperature	2.1.
T_{cs}	current sharing temperature	6.2.4.
T_{hs}	hot-spot temperature	7.2.
Thy_{fw}	free-wheeling thyristor	7.4.
T_r	reaction temperature	5.5.1.
T_w	heat source temperature	6.4.2.
T_λ	phase separation temperature	6.2.
t_{col}	collision time	13.2.2.
t_{nm}	normal zone recovery time	6.2.5.
t_f	film boiling onset time	6.2.3.
t_r	duration of reaction	5.5.1.
t_r	onset time, recovery time	6.2.3., 6.2.5., 6.4.3.
τ	electron scattering time	2.1.
Δt_d	triggering delay	7.3.
Δt_h	quench heater delay	7.3.
Δt_{sw}	power switch delay	7.3.
U	inner system energy	2.2.
U_1	voltage between layers	7.4.
U_m	maximum voltage in case of quenching	7.2.
μ	displacement	12.3.
μ_H	Hall voltage	13.2.2.
V	volume	2.2.
V, V_0	scalar magnetic potential	8.3., 9.2., 12.2.
V^*	conjugate scalar magnetic potential	12.2.

Symbol	Meaning	Chapter
VS, vs	integrated coil voltage	13.2.1.
v	electron velocity	2.1.
v	quench propagating velocity	7.2.
v_g	mass velocity	6.2.2.
v_{He}	helium coolant velocity	6.1.
v_s	superfluid velocity component of He II	7.2.
v_n	normal velocity component of He II	7.2.
v_{tr}	transverse component of quench propagation velocity	7.2.
W	mechanical work	2.2.
W_p	pumping power	6.3.
w	width of composite matrix	2.3.2.
x	quality factor of He I	6.2.2.
x_{tt}	Martinelli factor	6.2.2.
Z	(surface) impedance	6.2.2.

1

INTRODUCTION AND SUMMARY

The present book on *Superconducting High Field Magnets* contains 13 chapters. It is intended to cover the relevant interdisciplinary fields of applied physics with the aim of transmitting to the professional reader the vast theoretical work and experimental data which should help him to design, build and operate superconducting magnets and magnet systems.

Chapter 2, 'Basic features of superconductivity,' starts with the experimental discovery of the phenome by K. Onnes and analyses the intriguing fundamental experiment of Meissner and Ochsenfeld, which triggered vast theoretical and experimental activity with the aim of understanding and explaining superconductivity. The two-fluid thermodynamic theory and the intuitive but non-rigorous approach of the London brothers are presented, and this leads to the rigorous BCS theory, fully explaining the phenomenon. Basic parameters like critical magnetic fields and temperature, superconducting screening current, elementary quantized fluxoid, penetration depth and pinning forces are dealt with and the particular cases of Type I and Type II superconductors are given. High temperature superconductors are not dealt with.

Chapter 3, 'Flux-jumping in superconducting filaments and composites; stability criteria,' derives the stability conditions for superconducting filaments and composites in order to prevent flux-jumping or transitions into the normal conducting state. The main adiabatic and self-field stability criteria are recalled, which leads to the nowadays generally accepted limitations of the filament diameter, the need to incorporate stabilizing elements of high electric and thermal conductivity like copper or aluminium, of high resistivity barriers to reduce interstrand currents, and of filament twisting.

Chapter 4, 'Magnetization and AC losses in superconductors' analyses losses in filaments, composites and cables when subjected to time varying magnetic fields, which are also encountered in dc-operated magnets during excitation and deexitation. The subdivision of losses into magnetization and eddy type losses in filaments, composites and cables is given, whereby the derived expressions indicate ways of reducing them. In general reduced losses will positively influence the layout of the magnet cooling system, making it simpler and more efficient. Measuring techniques to determine the time-dependent losses are described.

Chapter 5, 'Manufacturing and application of advanced superconductors,' deals predominantly with the two main low temperature superconductors NbTi and Nb_3Sn. After enumerating the main requirements to be met, the more or less standardized manufacturing process used today of fine filamentary NbTi wires is described. The influence of certain parameters like the alloy composition, heat treatment and others is discussed, and the improved performance of superfluid He II–cooled NbTi superconductors is stressed.

First results on NbTi conductors, produced by a novel artifical pinning centre method, are presented. The second technical superconductor Nb_3Sn belonging to the intermetallic A15 compound group is extensively discussed. Different manufacturing methods such as the bronze route, the internal Sn method, including the tin tube source or TTS-variant, the jelly roll, and the Nb powder method are described and compared. The influence of He II cooling is reported and promising results are mentioned about Ta-doped Nb_3Sn conductors when operated in high field regions beyond 15 T. The sensitivity of brittle reacted Nb_3Sn wires and cables to longitudinal and transverse strain and stress is discussed. The chapter concludes with a brief review of other potential high field superconductors such as Nb_3Al, Chevrel-phase, and oxide based sheathed Bi-tapes, cooled to liquid helium temperatures.

The extensive Chapter 6 deals with the 'Cooling of superconducting high field magnets'. Identifying and then classifying the heat losses into steady state and short duration or transient ones, the main cooling modes are analysed: pool boiling He I cooling, forced flow two-phase and supercritical He I cooling, and the more recent cooling mode with superfluid He II. A distinction is made between heat transport through a medium, coolant or conductor, and heat transfer from conductor to coolant, both governed by the heat equation. Parameters in the form of experimental data to the heat equation are presented, notably heat transfer coefficients. Stability criteria are derived resulting from particular solutions to the heat equation and are known as full and partial cryostabilization and the minimum propagating normal zone cases. Analytical solutions to the heat equation for time-dependent energy disturbances are also given, whereby transient heat transfer coefficients by an order of magnitude higher than steady state figures between 300 and 2000 W m^{-2} K^{-1} have been measured.

Special attention is given to superfluid He II cooling. The thermo- and hydrodynamic properties of this two-fluid component liquid are described and the experimentally established Gorter–Mellink heat transport equation is presented. In a certain way heat transport in He II is similar to electron or current transport in superconductors. In He II heat is transported over long distances at minimum temperature gradients. Due to the high diffusivity, large volumes of He II contribute to the cooling of spot-like heat sources. Steady state and transient heat transfer coefficients and recent data on steady state heat transfer from insulated cables immersed in He II are presented. Finally the principle of the Claudet bath is explained, providing superfluid He II at atmospheric pressure rather than at the inconveniently low saturation pressure of a few mbars.

Chapter 7, 'The quenching process in superconducting magnets and their protection,' describes the quenching mechanism and discusses ways and means of efficiently protecting individual and strings of superconducting magnets against damage or destruction. Bearing in mind the steadily increasing stored energies of superconducting magnets, efficient quench protection must be integrated at an early stage into the magnet design. Input parameters defining the quench spreading mechanism and the critical values to be computed like the maximum 'hot spot' winding temperature and the coil overvoltages are discussed. Magnet protection by external resistor discharge and the concept of self-protected magnets are discussed, whereby in the latter case the entire coil volume is artificially made normal-conducting after activation of the coil heaters by the quench detector and energy evacuation by high power cold or warm diodes. Computer codes for accurate simulation of the quenching process are described and the close agreement between computed and measured quench phenomena is demonstrated.

Since the present book deals mainly with superconducting magnets for particle accelerators it seemed natural to briefly introduce them in Chapter 8, 'Transverse beam dynamics in circular accelerators for high energy physics'. Starting with the charged particle motion in these machines, the function of dipole, quadrupole and higher multipole magnets is described, with emphasis on strong-focusing accelerators. Transfer matrices for magnet elements are given and the phase space invariant, beam stability criteria and the dispersion theory are discussed. Transfer beam or beam optics requirements impose strict tolerances on the field quality, geometry and other magnet parameters.

In Chapter 9, 'Winding configurations for superconducting accelerator magnets,' the principal realistic coil configurations for superconducting dipole and quadrupole magnets are presented. Two-dimensional field expressions are derived from a scalar or vector potential and presented in the form of normal and skew field harmonics which are also responsible for part of the magnetic field errors. Constant current density intersecting ellipse—and concentric, circular or $\cos \theta$—like winding configurations are treated and expressions for the magnetic field in the different regions are presented. Expressions for accurate end field computation are given and the linear field error theory due to magnet tolerances is presented. Direct magnetic field computer codes will in general not provide winding configurations of the required field quality; to this end inverse codes are used, which will optimize winding configurations yielding almost error-free fields while respecting a number of imposed limititations and design constraints.

Chapter 10, 'Mechanical computations and design principles for superconducting magnets,' specifies the requirements to be satisfied by a safe and sound mechanical design. The mechanical forces and stresses in the transverse plane and the longitudinal end forces are computed. The commonly used hybrid structure is analysed and the function and contribution of collars, (split) iron yoke and external clamping structures to the winding precompression at different temperature and excitation levels explained. Some successfully built high field magnets in the 5–13 T field range are presented. The application and great utility of the computer codes Ansys and Castem for complex mechanical structure analysis are discussed. Among other items, insulation problems are presented. The successfully developed 'wind and react' technology for Nb_3Sn-wound high field magnets is explained. The chapter concludes by discussing mechanical design prospects for the next generation of superconducting accelerator magnets in the 15–16 T field range, whereby the multicurrent, multiblock coil design of distributed, reduced stresses is considered as a promising alternative.

In Chapter 11, 'Cooling of large accelerator magnet systems,' the heat inleaks and losses are enumerated. Three cooling systems, based on two-fluid cooling by heat exchange are described: the Hera, SSC and LHC systems. In the first two cases supercritical plus two-phase He I are used; for LHC, cooling with superfluid He II at atmospheric and saturation pressure is foreseen. The LHC collider magnet powering and protection schemes are also described.

Chapter 12, 'High field solenoids and detector magnets,' deals with the second large group of solenoidal magnets with special emphasis on high field NMR and NMI magnets and on large volume solenoids, integrated into huge particle detectors for high energy physics experiments. Expressions are given for the on and off axis field components in solenoidal winding configurations and the related forces and stresses are computed. The principle of nuclear magnetic resonance is briefly treated. The required field uniformity, design and manufacturing criteria for NMR magnets up to 25 T are given, and examples of existing and future detector solenoids for the LHC are discussed.

Chapter 13, 'Magnetic measurements of superconducting magnets,' concludes the book. The field range and measuring errors of the main field measuring devices such as rotating or pulsed coils, Hall probes and of NMR devices, are compared. The principle and the components of rotating coil measuring systems are discussed and the means to minimize the measuring errors are presented. The principle of Hall-probe field measurements is explained, the need for current and temperature stabilization is stressed, and manufacturing problems are addressed.

2

BASIC FEATURES OF SUPERCONDUCTIVITY

2.1 Discovery of superconductivity

The unexpected discovery of superconductivity by H.K. Onnes in 1911 in Leyden was due to research and development activities in two domains: the liquefaction of gases, leading in 1908 to the liquefaction of helium, and to the investigation of the electric conductivity of metals, which Onnes intended to explore at temperatures approaching absolute zero [1, 2]. According to Mathiessen's rule the total metal resistivity $\rho[\Omega m]$ consists of a temperature-dependent, dynamic component ρ_d (T) due to the scattering of electrons of thermal origin when subjected to an electric field $E[Vm^{-1}]$ and of a residual static component ρ_0 due to lattice impurities:

$$\rho = \rho_d(T) + \rho_0. \tag{2.1}$$

With e[As], m_e, [Ws^{3}m^{-2}], n_0 and τ the electron charge, mass, volume density and scattering time, the dynamic resistivity is:

$$\rho_d(T) = \frac{E}{j} = \frac{2m_e}{n_0 e^2 \tau} = \frac{E}{n_0 e v}. \tag{2.2}$$

Onnes had chosen distilled, purified mercury of very low ρ_0 and expected to find $\rho_d \rightarrow 0$ as $T \rightarrow 0$ [K]. The result of his experiment is shown in Fig. 2.1. At a critical temperature $T_{c0} = 4.15$K mercury became superconductive, the term given to the phenomenon by Onnes himself. He determined the residual resistivity to $< 5.10^{-25}[\Omega m]$ from the decay of the persistent currents in the sample. It was soon established that T_{c0} does not depend upon the electric field E, only slightly upon pressure p [Pa], but strongly on the magnetic field (or induction) H[Am^{-1}]. To any temperature $T < T_{c0}$ there is associated a critical magnetic field $H_c(T)$ according to Fig. 2.2:

$$H_c = H_{c0} \left\{ 1 - \left(\frac{T}{T_{c0}} \right)^2 \right\}, \tag{2.3}$$

which determines a transition or critical magnetic field H_{c0} beyond which no superconductive state exists. Table 2.1 shows T_{c0} and B_{c0} for certain pure metals.

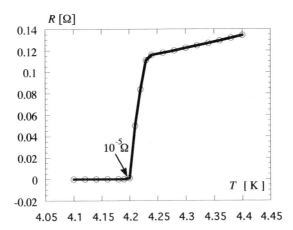

FIG. 2.1. The experiment of Kamerlingh Onnes.

Table 2.1 T_{c0} and B_{c0} values for some pure metals, Type I super-conductors

Metal	Nb	Pb	Ta	Hg	Sn	In	Ag
$T_{c0}[K]$	9.45	7.2	4.45	4.15	3.72	3.4	1.2
$B_{c0}[T]$	0.198	0.0803	0.083	0.0412	0.0309	0.029	0.0078

Equation (2.3) was verified by an experiment with a needle-shaped sample of a superconductor with a pickup coil, placed in an external parallel field and cooled to $T < T_c0$. As long as $H < H_c(T)$ superconductive screening currents keep H out of the superconductor. At $H = H_c$ they collapse, the magnetic field penetrates the superconductor and a signal appears across the pickup coil. So far superconductivity could be presented as a limiting case of normal conductivity with $\rho = 0$, since

$$\text{rot } \vec{E} = \text{rot } \vec{j} \ \rho = -\frac{\partial}{\partial t}\mu_0 H = 0 \qquad (2.4)$$

or $H = $ const. within the superconductor.

If the sample is first exposed to the external field and *then* cooled to $T < T_c(H)$ one would not expect any signal when crossing T_c, as the field should remain unchanged or frozen. Fortunately for the understanding of superconductivity, Meissner and Ochsenfeld [3] performed this experiment in 1933. Surprisingly, when crossing T_c a signal appeared, corresponding to the total expulsion of the magnetic flux in the sample, as shown in Fig. 2.3. At T_c a superconductive line current $I_1[\text{Am}^{-1}] = H_c$ is created. Both experiments confirmed the entire reversibility of the $I_1(H)$ behaviour as shown in Fig. 2.3a. If superconductivity could be explained by extending the metallic conduction

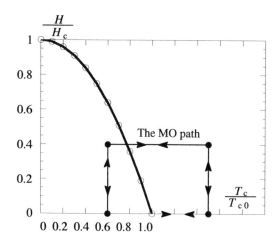

FIG. 2.2. The $H(T)$ diagram of Type I superconductors.

theory of now deactivated electrons to $\rho_d = 0$ one would logically expect an $I_1 = f(H)$ behaviour according to Fig. 2.3b. The Meissner–Ochsenfeld phenomenon is represented by the rectangular path in the $H(T)$ diagram in Fig. 2.2 and explained in Fig. 2.4. Independently of the direction of crossing the superconductive–normal conduction phase separation line, a signal appears across the pickup coil, induced by field expulsion when reducing $T < T_c(H)$ or by field penetration into the superconductor for $T > T_c(H)$. The screening current I_1 flows in a thin outside layer of the superconductor. Materials exhibiting this effect are called **Type I** or **soft superconductors**; according to Table 2.1 soft superconductors are mainly certain pure metals.

To explain superconductivity, other ways had to be investigated such as the thermodynamic phase change of the electron gas at T_c or the association of the attractive forces exhibiting electrons into molecules. Both ways will briefly be described as they lead to the BCS theory of Bardeen, Cooper, and Schrieffer, a detailed quantum physics theory explaining superconductivity.

2.2 The two-fluid thermodynamic theory

Gorter and Kasimir [4] developed in 1934 a thermodynamic, two-fluid model of normal conducting and superconductive electrons which contributed to a better understanding of superconductivity. In analogy with the gaseous, liquid and solid phases of atoms and molecules, depending upon pressure ρ [Pa] and temperature T, the $H(T)$ curve according to Fig. 2.2 was interpreted as a borderline between two systems; the inner, condensed, superconductive one, where the dc current is transported without resistance, and the outer, normal conducting region of resistive electron conduction. The two phase variables are T and H; along the separation line normal conductance and superconduc-

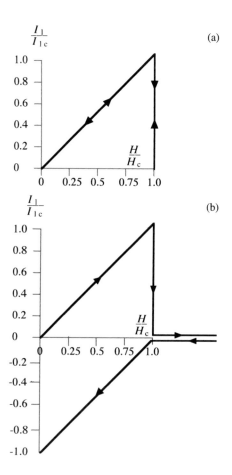

FIG. 2.3. Behaviour of a superconductor (a) and of a normal conductor (b) with resistivities $\rho = 0$.

tivity coexist in the sample. Thermodynamics distinguishes two forms of energy, heat
$Q[\text{Ws}]$ and mechanical work $W[\text{Ws}]$, which are both part of the inner energy of the
system :

$$U = Q + W \tag{2.5}$$

with

$$Q = TS \tag{2.6}$$

where $S[\mathrm{WsK^{-1}}]$ is the entropy of the system. Defining now a composite free energy F

$$F = U - TS \tag{2.7}$$

the difference between two states F_2 and F_1 is

$$dF = dU - T dS - S dT. \tag{2.8}$$

For an isothermal process with $T = \mathrm{const.}, dT = 0$

$$dF = dU - dQ = dW. \tag{2.9}$$

The difference between two free energy states is equal to the mechanical work.

Applying the two-fluid theory to phase transitions in superconductors, the following experiment has been performed: at constant temperature $T < T_{c0}$ a needle-shaped superconductive sample of volume $V [\mathrm{m^3}]$ was placed into a collinear outside magnetic field H with a sizeable gradient $-\partial H/\partial x$. Due to the interaction between the screening current I_1 and the field gradient $-\partial H/\partial x$ the sample is submitted to a repulsive force f

$$f = -\mu_0 V I_1 \frac{\partial H}{\partial x}. \tag{2.10}$$

To reach the critical field H_c the sample must be pushed backwards; the equivalent mechanical work is:

$$W = \int f dx = \int_0^{H_e} \mu_0 V H dH = \mu_0 V \frac{H_c^2}{2}. \tag{2.11}$$

Due to $T = \mathrm{const.}, dF = W$ and

$$F_{sc}(T, H) = F_{sc}(T, 0) + \mu_0 V \frac{H^2}{2}. \tag{2.12}$$

At $H = H_c$ the sample goes normal and due to the reversibility of the process no additional work is required; eqn (2.12) is then:

$$F_{sc}(T, H_c) = F_{sc}(T, 0) + \mu_0 V \frac{H_c^2}{2} \tag{2.13}$$
$$= F_{nc}(T, H_c) = F_{nc}(T, 0)$$

and

$$F_{\mathrm{nc}}(T, 0) - F_{\mathrm{sc}}(T, 0) = \mu_0 V \frac{[H_{\mathrm{c}}(T)]^2}{2}. \tag{2.14}$$

Here 'sc' and 'nc' stand for superconductive and normal conducting states. In zero field the superconducting phase has a **lower** free energy and is therefore thermodynamically stable. By raising H to H_c the screening current effect raises the free energy in the superconductor until it becomes normal conductive. From eqn (2.7) one obtains for the entropy at $H = \mathrm{const.}$:

$$S = -\left(\frac{\partial F}{\partial T}\right) H. \tag{2.15}$$

and for the electron contribution to the specific heat C_{e} :

$$C_{\mathrm{e}} = T \frac{\mathrm{d}S}{\mathrm{d}T}. \tag{2.16}$$

Applied to eqn (2.14) one obtains for $H = 0$:

$$S_{\mathrm{nc}} - S_{\mathrm{sc}} = -\mu_0 V H_{\mathrm{c}} \frac{\partial}{\partial T} H_{\mathrm{c}}(T). \tag{2.17}$$

and for the difference between the corresponding specific heats:

$$C_{\mathrm{e,nc}} - C_{\mathrm{e,sc}} = -\mu_0 V T \frac{\partial}{\partial T} \left\{ H_{\mathrm{c}}(T) \frac{\partial}{\partial T} H_{\mathrm{c}}(T) \right\}. \tag{2.18}$$

Both curves have zero slope at $T = 0$ which is not the case for conduction electrons, treated as a Fermi gas; their entropy S_{nc} and specific heat $C_{\mathrm{e,nc}}$ rise linearly with temperature, while eqns (2.17) and (2.18) suggest an exponential rise in the range $0 < T < T_{c0}$ as verified by experiment. The condensation effect in a superconductor is thus a gradual process like freezing of water: at T_{c0} the first superconducting charge carriers of the new fluid appear; in the interval $0 < T < T_{c0}$ normal conducting and superconducting fluids coexist and only at $T = 0$ will the last normal conducting electrons be transformed into superconducting ones.

2.3 Skin effect in superconductors

The field exclusion from the bulk of a superconductor has so far been attributed to a surface current of linear current density $I_1 = H$. Physically, however, the superconductive current must have a **finite penetration depth**. This requirement and the field expulsion demonstrated by the MO effect were explained by the intuitive skin effect theory of the London brothers in 1935 [5] , the work of Landau and Lifschitz [6,7], and finally by the rigorous quantum physics theory of Bardeen, Cooper, and Schrieffer in 1952 [8].

To understand this evolution, the skin effect in normal conducting metallic conductors is recalled. Introducing a vector-potential $\vec{A}$ one obtains from Maxwell's equations:

$$\vec{B} = \mu_0 H = \text{rot } \vec{A} \ . \tag{2.19}$$

$$\frac{\partial}{\partial t} \vec{B} = \text{rot } \vec{E} = \text{rot}\rho_n \ \vec{j} = \rho_n \text{rot } \vec{H} \tag{2.20}$$

$$\text{rot}\left[\text{rot } \vec{H}\right] = -\frac{\mu_0}{\rho_n} \frac{\delta}{\delta t} \vec{H} \ . \tag{2.21}$$

With the solution for an alternating field $H_m\exp(i\omega t)$ eqn (2.21) yields:

$$H(x,t) = H_m\exp\left(-\frac{x}{\delta}\right)\exp\left[i\left(\omega t - \frac{x}{\delta}\right)\right] \tag{2.22}$$

with the skin depth, dissipation resistance, and inductance

$$\delta = \sqrt{\frac{2\rho_n}{\omega\mu_0}} \tag{2.23}$$

$$R = \sqrt{\frac{\omega\mu_0\rho_n}{2}} \tag{2.24}$$

$$L = \frac{\mu_0\delta}{2} . \tag{2.25}$$

For a finite frequency and $\rho_n \to 0$, eqns (2.23)–(2.25) would yield $\delta = R = L = 0$, contrary to experimental results. In 1933 Becker, Heller, and Sauter introduced the unscattered electron mass m_e

$$\frac{\delta}{\delta t}m_e \ \vec{v} = e \ \vec{E} \tag{2.26}$$

$$\frac{\delta}{\delta t}\left(ne \ \vec{v}\right) = \frac{\delta}{\delta t} \vec{j} \tag{2.27}$$

leading to:

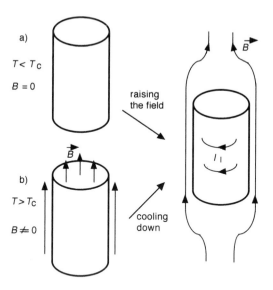

FIG. 2.4. Explanation of the Meissner–Ochsenfeld effect by the behaviour of a lead
cylinder in a magnetic field.

$$\text{rot}\frac{\delta}{\delta t}\,\vec{j} = \text{rot}\left[\text{rot}\frac{\delta}{\delta t}\,\vec{H}\right] = -\frac{\mu_0 n e^2}{m}\frac{\delta\,\vec{H}}{\delta t} \qquad (2.28)$$

with the solution

$$H(x,t) = f(t)H_m\exp(-x/\Lambda) + H_0(x) \qquad (2.28a) \qquad\qquad (2.29)$$

$$\Lambda = \sqrt{\frac{m_e}{\mu_0 n e^2}}.$$

Λ is the penetration depth for the dc case $f(t) = 1$—see Fig. 2.5. However, this
approach admitted a frozen internal magnetic field $H_0(x)$ in contradiction with the MO
experiment. In this situation F. and H. London proposed with genuine intuition but without proof the following relation for the superconductive current density part j_{sc}:

$$\text{rot}\,\vec{j}_{sc} = \text{rot}\left[\text{rot}\,\vec{H}\right] = -\frac{1}{\Lambda^2}\,\vec{H} \qquad (2.30)$$

with the solution

$$H(x) = H_m \exp(-x/\Lambda). \tag{2.31}$$

This solution is now in agreement with the MO experiment since no internal field is admitted. The penetration depth of the superconductive current is $\Lambda \sim 10^{-7}$ m.

The BCS quantum theory provided the rigorous solution by showing that electrons with opposite spin and momentum can be bound together into **Cooper pairs** of mass $m_s = 2m_e$ and charge $q_s = 2e$. The Cooper pairs form the superconductive fluid mentioned in Section 2.2. Equation (2.29) is thus modified to

$$\Lambda_{bcs} = \sqrt{\frac{m_s}{\mu_0 n_s q_s^2}}. \tag{2.32}$$

According to the BCS theory the normal conducting electron density decreases rapidly below $T < T_c/2$ since:

$$n = n_0 \exp(-2T_c/T). \tag{2.33}$$

At $T_c/2$ only 1.8% of electrons are not yet paired.

The BCS theory is also important for RF superconductivity; if one introduces into eqn (2.31) a time-dependent field

$$H(x, t) = H_m \exp(-x/\Lambda) \exp(i\omega t) \tag{2.34}$$

the electric field E will act on the normal conducting electrons. The related surface impedance Z is then:

$$Z = R + iL\omega = \frac{\mu_0 \Lambda^3 \omega^2}{2\rho_n} \exp(2T_c/T) + i\omega\mu_0\Lambda. \tag{2.35}$$

Relation (2.35) is of considerable importance to superconductive RF accelerating cavities since it influences the choice of their superconductive and normal conducting layers.

2.4   Flux quantization

Extending our present knowledge to a thin superconductive cylindrical tube according to Fig. 2.5 exposed to a coaxial magnetic induction B and cooling the tube to $T_c(B)$ and below, a magnetic flux Φ[Vs] will be trapped inside the hole. Two circumferential superconductive currents will flow within the penetration depth on the inner and outer tube surface excluding the magnetic induction from the bulk of the wall. When the external induction is removed only the outer screening current will disappear, while the inner one will continue to flow and shield the trapped flux Φ. F. London postulated in 1950 that this flux must be **quantized**:

$$\Phi = m\Phi_0 \tag{2.36}$$

with the **elementary fluxoid** equal to

$$\Phi_0 = \frac{h}{q_s} = \frac{h}{2e} = 2 \times 10^{-15} \text{Vs} \tag{2.37}$$

where h is Planck's constant. Relation (2.37) has been confirmed by the BCS theory [8] and verified with considerable precision by the experiments of Doll and Näbaur [9] and of Deaver and Fairbank [10] in 1961, using thin lead and tin tubes.

So far only 'ideal' or Type I superconductors have been considered. According to Table 2.1 their critical magnetic inductions B_{c0} are rather low, below the 1.5–2.0 T range normally obtained in classical, iron contour dominated magnets. In fact when measuring the magnetic flux difference between a normal and superconductive sample such as lead of identical length and cross-section one obtains the curves shown in Fig. 2.6a. When lead is replaced by a Pb-In alloy one obtains the curve shown in Fig. 2.6b. The alloy, a **Type II superconductor**, shows complete field exclusion below the lower critical field H_{c1}; when the field is increased to the upper critical field H_{c2} the sample remains superconductive; the surfaces s_1 and s_2 shown in Figs 2.6a and 2.6b are equal. This means that the work required to bring the Pb-In sample into the field H_{c2} is the same as the work to bring the Pb sample to its H_c.

This behaviour is explained by the splitting, or breaking up, of the sample into normal and superconducting zones. The flux is concentrated in tubular isles, each carrying an elementary fluxoid Φ_0 located in the normal zones as shown in Fig. 2.7. This model was proposed in 1952 by Abrikosov [11] and verified in 1968 by Träuble and Essmann [12]. Each of the tubular flux isles thus **punches** a normal conducting hole into the sample. For a better understanding of mixed state Type II superconductors one has to compute the mechanical work experienced by a needle sample in a collinear external field H with a gradient $-\partial H/\partial x$, similar to Section 2.2. The resulting forces acting on a flux tube model of length l, cross-section a, and a line current I_1, brought into the field H of a needle-shaped sample of cross-section s, are:

$$f = f_t + f_s = l\,[\mu_0 s H - \Phi_0]\,\frac{\partial H}{\partial x} \tag{2.38}$$

and the work to bring a sample with one flux tube into the field H is:

$$W = W_s - \Delta W = l\left(s\mu_0\frac{H^2}{2} - \Phi_0 H\right). \tag{2.39}$$

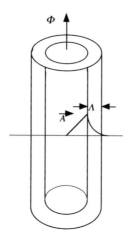

FIG. 2.5. A thin superconductive cylinder at $T < T_0$ after the external field has been removed.

Introducing $\xi[m]$ for the radius of the normal conducting hole or the coherence length as shown in Fig. 2.7b, one obtains from eqns (2.14) and (2.39) that the energy loss ΔW due to one flux-tube is compensated by the gain due to the fluxoid contribution:

$$\Delta W = \xi^2 \pi l \frac{\mu_0 H_c^2}{2} = l H_{c1} \Phi_0. \tag{2.40}$$

For large enough ξ, H_{c1}, a quadratic function of ξ will exceed H_c and the super-conductive state will be destroyed before a single fluxoid is formed; we have a **Type I** superconductor. At $H = H_{c1}$ the fluxoid magnetic diameter is assumed to 2Λ such that:

$$\Phi_0 = \mu_0 \Lambda^2 \pi H_{c1} \tag{2.41}$$

$$H_{c1} = \frac{\xi}{\sqrt{2\Lambda}} H_c \tag{2.42}$$

Equation (2.42) states that for $\xi > \sqrt{2\Lambda}$ the superconductor is of Type I and for $\xi < \sqrt{2\Lambda}$ of Type II. The upper critical field H_{c2} can be calculated in a similar way by assuming that all normal conducting cores of the flux-tubes are in contact:

$$\Phi_0 = \xi^2 \pi \mu_0 H_{c2} \tag{2.43}$$

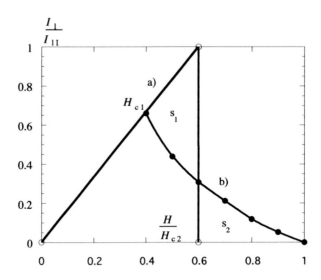

FIG. 2.6. $I_1/I_{11} = f\,(H/H_{c2})$ diagrams for (a) Type I and (b) Type II superconductors.

$$H_{c2} = \frac{1}{\sqrt{2}}\frac{\Lambda}{\xi}H_c; \quad H_{c1}\cdot H_{c2} = H_c^2. \tag{2.44}$$

The structure of the magnetic flux-lines in Type II superconductors up to H_{c2} is shown in Fig. 2.7 with the circulating supercurrents around each fluxoid Φ_0. In order to pass Φ_0 through the specimen, the density of the Cooper electron pairs, concentrated within the small coherence length ξ, will decrease just near the flux-line axis. This model was first proposed by A. Abrikosov [11] in 1957. The T-H phase diagram of Type II superconductors is shown in Fig. 2.8.

Can such a symmetric fluxoid lattice according to Fig. 2.7a with a uniform flux repartition meet our main requirement for a technical superconductor to carry a sizeable current density j_s in a magnetic field $H < H_{c2}$? Certainly *not*, as the application of eqn (2.21) yields $j_s = 0$ for B or $\Phi_0 = $ const. No net macroscopic current is allowed within such a **soft** Type II superconductor. In a symmetric fluxoid mesh according to Fig. 2.7a and 2.7b the superconductive vortices and their repulsive forces cancel out except in the outermost circumferential layer. Furthermore, if one tries to pass an increasing so called transport current I_t through the conductor, the accompaning circular magnetic field H_t would at $H_t > H_{c1}$ start pushing the flux-tubes inwards, destroying the outer ones.

A **non-uniform** fluxoid pattern is thus needed for a finite j_s; due to the spatial gradient $\partial \vec{H}/\partial s$ the net repulsive force between fluxoids is then:

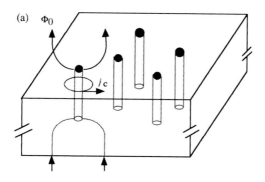

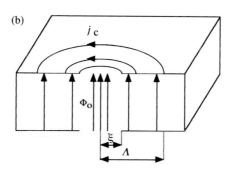

FIG. 2.7. (a) Elementary fluxoids Φ_0 located in the normal zones of a soft Type II superconductor. (b) Penetration depth Λ and coherence length ξ in a Type II superconductor (Figs 2.7a and 2.7b courtesy of Prof. W. Buckel, Germany).

$$f_s = \vec{H}\, \mu_0 \frac{d\vec{H}}{ds} = \mu_0 \left[\vec{H}\, x\mathrm{rot}\, \vec{H} \right] = \left[\vec{B}\, xj_s \right] \tag{2.45}$$

which is also the Lorentz force due to the macroscopic current density j_s.

Fortunately in hard Type II superconductors artificial structural defects of the elementary fluxoid size are built into the crystal lattice giving rise to a **pinning force** f_p [N], opposed to the Lorentz force and preventing movements of the fluxoids. A critical pinning force $\vec{f}_{pc}$ can be defined at $\vec{j}_{sc}$ and $\vec{B}$:

$$\vec{f}_{pc} = \left[\vec{B} \times \vec{j}_{sc} \right] \tag{2.46}$$

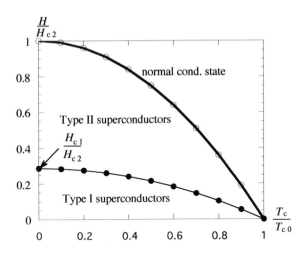

FIG. 2.8. Phase diagrams for Type I and Type II superconductors.

Table 2.2 B_{c2} and T_{c0} values of hard Type II superconductors

Superconductor	NbTi	Nb$_3$Sn	Nb$_3$Al	Nb$_3$(AlGe)	Chevrel phase
$B_{c2}[T]$	14–15	24–30	36–41	43.5	60
$T_{c0}[K]$	9.0	18.2	19.1	19.3	14.5

whereby $j_{sc} = f(B)$ at $j = j_{sc}$ flux-tubes begin to move, a resistive component appears, and the superconductor goes normal. One can now describe hard Type II superconductors by the model of Bean [13] and Kim [14]: when raising the field H_t around a superconductor according to Fig. 2.9 it will progressively penetrate its cross-section and supercurrents of critical density j_{sc} will flow in the penetration zone. As j_{sc} is inversely proportional to H or B as demonstrated by Anderson [15]

$$j_{sc} = \frac{j_0 B_0}{B + B_0}. \tag{2.47}$$

These supercurrents will soon occupy the entire cross-section and the magnetic field H will have fully penetrated the superconductor. At $H = H_{c2}$, $j_{sc} = 0$, the superconductor will go normal.

Table 2.2 gives the upper critical magnetic inductions B_{c2} and the T_{c0} values for a number of hard Type II superconductors.

Today NbTi and Nb$_3$Sn superconductors are almost exlusively used in high field magnets. Figure 2.10 shows the ultimate or critical pinning forces f_{pc} for a number of hard Type II superconductors. If high f_{pc} superconductors are certainly desired, their reliable manufacturing process still needs further development. However, the curves

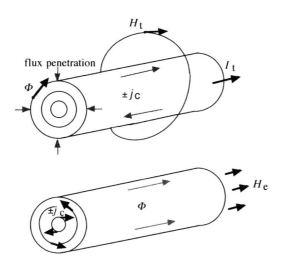

FIG. 2.9. Flux penetration into a hard Type II superconductor for $H > H_1$.

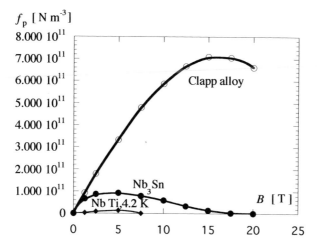

FIG. 2.10. The pinning force f_p of some hard Type II superconductors.

shown in Fig. 2.10 provide useful indications on the development of promising **APC**
(artificial pinning centre) superconductors for the future. Some of them will be briefly
discussed in a later chapter.

3

FLUX-JUMPING IN SUPERCONDUCTIVE FILAMENTS AND COMPOSITES; STABILITY CRITERIA

3.1 Introduction

The emergence of hard Type-II superconductors, once their behaviour and mechanism had been understood—see Chapter 2—raised considerable expectations as to their application in different domains. Increased performance and savings were expected from the use of high magnetic fields such as reduced energy losses in the excitation windings of electrical transformers and machines, smaller running costs and devices of reduced size, compared to 'classical' ones. When by the early 1960s the first technical superconductors became available in the form of NbTi wires of 100–250 mm diameter and of Nb_3Sn tapes of ~ 0.1 mm thickness, the obtained performances were disappointing; the test coils either failed at currents and magnetic fields well below the critical values or exhibited considerable training, a large number of quenches before approaching nominal performance.

This situation triggered significant theoretical and experimental work with the aim of understanding and eliminating the causes of the disappointing performance of coils, wound with these first hard Type II superconductors. Two lines of action which promised solutions had been followed: already in 1964 Chester [1,2] stressed the need for thin fine superconductive filaments by a simple calculation showing that a sudden 2×10^{-8} [W s m^{-3}] energy release in a 200 μmΦ NbTi wire with an assumed current density of 3×10^9 A m^{-1} would raise the wire temperature by 10 K and result in a quench. When subdividing this filament into 100 filaments of 20μmΦ the ΔT would be limited to only 0.3 K.

The need and importance of an additional stabilizing element of high electric and thermal conductivity had also been recognized. In fact the only coils attaining critical current and field values had fully cryostabilized superconductors—see Chapter 6—with enough Cu or Al to carry the nominal current in case of a perturbation.

In this context the fundamental works of the Rutherford Laboratories Superconducting Magnet Group (M.N.Wilson, C.R.Walters, J.D.Lewin and P.F.Smith), Hancox [3,4] and Turck and Duchateau [5,6] is mentioned. It led to the development of fine filamentary, twisted and mainly Cu-stabilized wires and composites as basic elements in multi-strand superconductive cables or compact conductors. This fundamental work had been completed within a decade. Since then conductor and cable performances are being steadily improved and fine filamentary Nb3Sn composites have been developed for currents up to 10^5 A or alternatively for inductions in the 20 T range.

3.2 Stability criteria for superconductive filaments and composites

It has been shown in Chapter 2 that the magnetic flux penetrates the cross-section of a hard Type II superconductor in the form of quantized elementary fluxoids, surrounded by vortices of supercurrents at critical current density j_c which in turn depends upon the local magnetic field $j_c = f(H)$. According to Fig. 3.1 two cases are distinguished:

a) The superconductor is placed in an external magnetic field H_e but carries no transport current I_t; equal and opposite screening currents of density $\pm j_c$ flow in the plane orthogonal to H_e. Except for specific screening devices the case shown in Fig. 3.1a is of little practical importance.

b) The superconductor is exposed to H_e and carries a transport current I_t. As shown in Fig. 3.1b this is the situation in most superconducting coils. Superimposed screening and transport currents $\pm I_c$ and I_t will flow simultaneously, whereby their respective orientation will depend upon the location of the superconductor in H_e.

We shall now examine the conditions for the stable operation of superconductive filaments and composites, known as the **adiabatic** and **dynamic** stability criteria.

3.2.1 *Adiabatic stability*

The adiabatic stability criterion, which is also the most stringent one, considers the rectangular or circular superconductive filament to be thermally isolated during a perturbation and determines the limiting conditions under which the heat capacity or the specific heat per unit volume of the filament $\gamma C[\text{Wsm}^{-3}K]$ will absorb the released energy. One can imagine the following closed loop of events: an energy disturbance ΔQ_1 provokes a temperature increase ΔT which in return results in a critical current reduction $-\Delta j_c$ and a deeper penetration into the superconductor Δr, accompanied by a flux change $\Delta \Phi$. The result is a further energy increase ΔQ_2 and so on. Under certain conditions $\sum \Delta Q$ will ultimately lead to a transition into the normal state or to a quench.

For a superconductive slab of width $2a_f$ of specific heat γC, critical temperature margin $T_c - T_0$ and j_c placed in an external field H_e perpendicular to the slab width, the stability condition is:

$$a_f \leq \frac{\sqrt{3}\,(T_c - T_0)\,\gamma C}{\dfrac{\mu_0}{j_c}}. \tag{3.1}$$

For a better comparison one introduces the stability parameter β:

$$\beta = \frac{\mu_0 j_c^2 a_f^2}{\gamma C (T_c - T_0)} \leq 3. \tag{3.2}$$

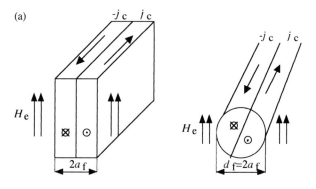

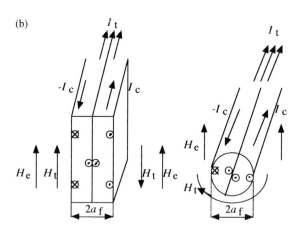

FIG. 3.1. (a) Superconductive slab and wire in external field H_e carrying only $\pm I_c$; (b) superconducting slab and wire in external field carrying I_t and $\pm I_c$.

More rigorous adiabatic stability theories of Bean and Swartz [7] resulted in $\beta \leq \pi^2/4$. For a cylindrical superconductor of radius r_f the stability parameter is $\beta \leq 2.7$. Let us now calculate the allowed a_f for a Nb-47% Ti filament at $B_e = 5T$, with the following parameters: $j_c = 2.8 \times 10^9 [\text{Am}^{-2}]$, $T_c - T_0 = 7.0 - 4.2 = 2.8$ K and $\gamma C = 7 \times 10^3 [\text{Wsm}^{-3}\text{K}^{-1}]$; the result is $a_f < 77\mu$m. A similar calculation for a high j_c fine filamentary Nb$_3$Sn conductor at $B_e = 10T$, $T_c - T_0 = 10 - 4.2 = 5.8\text{K}(\gamma C)_{\text{av}} = 7 \times 10^3 [\text{Wsm}^{-3}K^{-1}$ and $j_c = 4 \times 10^9 [\text{Am}^{-2}]$ in the Nb$_3$Sn part yields $a_f = 74\mu$m.

At low inductions of $B_e \approx 1T$ the criterion yields $a_{f\,\text{Nb}_3\text{Sn}} \leq 42\mu$m, $a_{f\,\text{NbTi}} \leq 50\mu$m. Advanced technical superconductors meet these conditions; the need for even smaller filament diameters of a few μm is determined by other requirements like small magnetization and eddy current losses—see Chapter 4—and/or small field distortions due to magnetization currents—see Chapter 9.

An interesting contribution to adiabatic stability is the **partial** flux-jumping theory elaborated by Hancox [3a,4a]. In a superconductor carrying a transport and magnetization currents I_t and $\pm I_c$ at a critical current density j_{c1} partial flux-jumping can under certain conditions lead to subsequent quenching of the magnetization currents, leaving only the transport current at a lower $j_{c2} < j_{c1}$. The condition for partial flux-jumping is

$$Q = \frac{\mu_0 a_f^2}{6} \left(j_{c1}^2 - j_{c2}^2 + 6 j_{c2}^2 \ln \frac{j_{c1}}{j_{c2}} \right) \leq \int_{T_1}^{T_2} \gamma C dT. \tag{3.3}$$

In the limiting case when $j_{c1} \to j_{c2}$ one obtains eqn (3.1) for the adiabatic stability of a superconducting slab carrying the transport current over its entire cross-section.

3.2.2 Dynamic stability

The fragmentation of superconductors into fine filaments required their bundling or grouping into larger elements, called composites, and wires in order to obtain reasonable currents. The filaments are embedded in a matrix of high electric and thermal conductivity; the interplay between filaments and matrix had to be analysed and understood.

Dynamic stability criteria and the requirement for filament twisting had been derived assuming that any local heat pulse will be absorbed and rapidly distributed over the cross-section of the composite due to the high thermal diffusivity D_{th} of the matrix and to filament coupling by the high electric conductivity of the matrix. The dynamic stability will be computed for a composite of radius a [m] with filaments of radius r_f [m], matrix thermal conductivity k[W m^{-1}K^{-1}] and electric conductivity ρ^{-1}[Ω^{-1}m^{-1}]. The magnetic induction B_e and the transport current I_t are assumed to be linked by the linear (load line) relation:

$$K = \frac{B_e}{I_t}[\text{TA}^{-1}]. \tag{3.4}$$

When raising I_t the outer filaments will saturate first as the combined H_e and H_t penetrate into the composite, see Fig. 3.1b. As will be shown later, filament twisting practically cancels the effect of penetration of H_e, but has no influence on the penetration of H_t. By saturating the outer shell an electric field of $E \sim 2 \times 10^4$Vm^{-1} [8] is created causing heat dissipation which can lead to a catastrophic flux-jump; this is the **self-field instability**. The task of the low resistivity matrix is to increase D_{th}[ms^{-2}] and reduce the magnetic diffusivity D_m of the composite, thus preventing flux-jumping and ensuring dynamic stability. Following the approach of Turck and Duchateau [5,6] the dynamic stability of a composite can be calculated as follows: the average $(\gamma C)_{av}$, k_{av}, and resistivity ρ_{av} are determined by taking the volume part λ of the superconductor into account:

$$(\gamma C)_{av} = \lambda(\gamma C)_{sc} + (1 - \lambda)(\gamma C)_{Cu} \tag{3.5}$$

$$k_{av} = k_{Cu}\frac{1 - \sqrt{\lambda}}{1 + \lambda - \sqrt{\lambda}} \tag{3.6}$$

$$\rho_{av} = \rho_{Cu}(1 - \lambda). \tag{3.7}$$

Introducing further the temperature parameter

$$\tau_{c0} = \frac{-j_{c0}}{\dfrac{\delta j}{\delta T}} \tag{3.8}$$

the condition for dynamic stability is

$$a \leq \frac{\sqrt{\beta_c(\gamma C)_{av}\tau/\mu_0}}{\lambda j_c} \tag{3.9}$$

or

$$\beta_c \leq \frac{\mu_0(\lambda j_c)^2 a^2}{(\gamma C)_{av}\tau_{c0}}. \tag{3.10}$$

The influence of the cooling conditions is taken into account by the normalized heat coefficient

$$h = \frac{h_{co}a}{k_{av}} \tag{3.11}$$

with $h_{co} \approx 2 - 5 \times 10^4 [W\ m^{-2}K^{-1}]$ for liquid helium cooling of transient phenomena. The maximum permissible normalized transport current $I_t/I_c = i$ can be determined from Figs 3.2 and 3.3. One first computes h, β_c and $v = D_{th}/D_m$ using eqns (3.10), (3.11), (3.20) and (3.21). From Fig. 3.2 the maximum normalized degraded current $i_d = I/I_c$ due to saturation of the composite outer layer is found. i is then obtained from Fig. 3.3 for different parameters of K according to eqn (3.4). Calculations and experiments confirmed the improved performances obtained by simultaneously raising B_e and I_t.

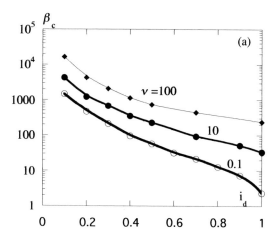

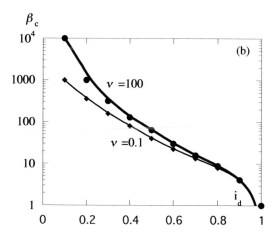

FIG. 3.2. Critical parameter β_c versus degraded current i_d for the cooling parameter (a) $h = 1$; (b) $h < 0.001$.

Wilson [9] has calculated the self-field stability parameter β_t in a similar way. Assuming that I_t flows within a progressively increasing outer ring of the composite between radii a and c:

$$\beta_t = \frac{\mu_0 \lambda^2 j_c^2 a^2}{(\gamma C)_{av}(T_c - T_0)} \leq \frac{1}{-0.5\ln\varepsilon - \dfrac{3}{8} + \dfrac{\varepsilon^2}{2} - \dfrac{\varepsilon^4}{8}} \tag{3.12}$$

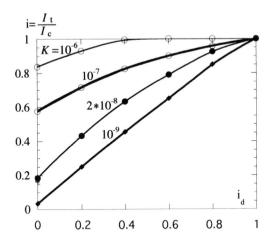

FIG. 3.3. Overall transport current versus degraded current i_d.

with

$$\varepsilon^2 = 1 - i = 1 - \frac{I_t}{I_c} = \frac{c^2}{a^2}.$$ (3.13)

For the **dynamic** stability of I_t in a composite Wilson finds:

$$\beta_{t \text{ dyn}} = \frac{\mu_0 \lambda^2 j_c^2 a^2}{(\gamma C)_{\text{av}}(T - T_0)} \leq \frac{1 + \dfrac{2\xi}{\alpha^2}}{-0.5 \ln \varepsilon - \dfrac{3}{8} + \dfrac{\varepsilon^2}{2} - \dfrac{\varepsilon^4}{8}}$$ (3.14)

with the cooling parameter

$$\xi = \frac{h_{c0}\mu_0(1 - \lambda)}{\rho_{\text{Cu}}(\gamma C)_{\text{av}}}$$ (3.15)

and $\alpha = f(\xi)$ according to Fig. 3.5.

For a stack of Cu-stabilized superconductive slabs of width $2a$ and the magnetic field H_e perpendicular to the slab width the dynamic criterion for the screening or magnetization currents is:

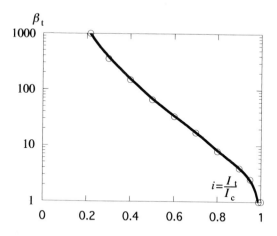

FIG. 3.4. Self-field parameter β_t depending upon i.

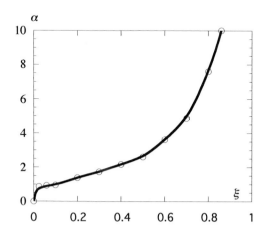

FIG. 3.5. α-dependence upon the cooling parameter ξ for calculating the dynamic stability of I_t.

$$\beta_{s\ dyn} \leq 3\left(1 + \frac{4\xi}{\pi^2}\right). \tag{3.16}$$

When I_t is also present the stability criterion becomes:

$$\beta_{s\ dyn} \leq \frac{3}{1+3i^2}\left[1 + \frac{4\xi}{\pi^2}(1+i)^2\right]. \tag{3.17}$$

On several occasions the magnetic and thermal diffusivity D_m and D_{th} have been mentioned in relation to dynamic stability criteria. The diffusion equation defining the linear motion of heat, magnetic induction, or current density (electric field) in a superconductor or stabilizer is:

$$D_n\frac{d^2N}{dx^2} = \frac{\delta N}{\delta t} \tag{3.18}$$

with the solution for a slab of width $2a$ [10]:

$$T, B \text{ or } j_c = \sum_1^\infty A_n \sin\left(\frac{n\pi x}{2a}\right)\exp\left(\frac{D_n n^2 \pi^2 t}{4a^2}\right). \tag{3.19}$$

For D_{th} and D_m one obtains:

$$D_{th} = \frac{k}{\gamma C}; \quad D_m = \frac{\rho}{\mu_0} \tag{3.20}$$

and for the main time constant ($n = 1$)

$$\tau_1 = \frac{4a^2}{\pi^2 D_1}. \tag{3.21}$$

Superconductors exhibit high D_m and low D_{th} values, while stabilizing metals have high D_{th} and low D_m diffusivities in the proportion of $10^4 - 10^5$ in both cases.

3.2.3   Twistpitch and transposition

Twisting of superconductive filaments in composites and wires has been incorporated into the fabrication process of technical superconductors since the end of the 1960s. To explain the physical arguments, a slab model with adjacent filaments of diameter d_f, separated by a high conductivity matrix of width w, is shown in Fig. 3.6. Exposed to a perpendicular, **time varying** magnetic induction $\dot{B}_e$—which also appears in dc

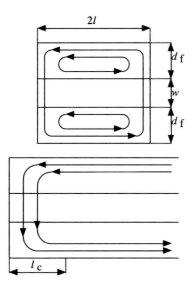

FIG. 3.6. Effect of twisted (top) and non-twisted (bottom) filaments.

magnets during excitation and deexcitation—current loops of j_c are induced, flowing in the filaments and across the resistive matrix. At a critical length l_c [m] the entire current I_c will flow through the matrix; l_c is given by:

$$I_c = \sqrt{\frac{2 j_c \lambda d_f \rho}{\dot{B}} \frac{w}{w + d_f}}. \tag{3.22}$$

For a circular composite the factor 2 has to be replaced by $\pi^2/4 = 2.47$. For $l \leq l_c$ only the fraction $(l/l_c)^2$ of the magnetization currents will cross the matrix and the ohmic losses will be reduced—see Fig. 3.6. At $l > l_c$ or in the case of non-twisted composites $l_c = \infty$ the transverse currents will flow at the composite ends, as shown in Fig. 3.6. The **twistpitch length** corresponding to a full period amounts to $4l$ [m].

Another reason for filament twisting is the non-uniformity of $\dot{B}$ over the composite length in a superconductive winding, which may well attain several km; j_c gradients in the screening currents would otherwise drive transverse currents across the matrix. In that case twisting will subdivide the filaments into loops of length $2l$, each loop carrying a j_c matched to the local induction B. Twistpitch length in the technical superconductors NbTi and Nb₃Sn amount to 0.016m $< 4l <$ 0.025m. Calculating now I_c for a composite with $\rho = 2 \times 10^{-10}\Omega$m, $\dot{B} = 0.004$Ts^{-1}, $j_c = 3 \times 10^9$Am^{-2}, $\lambda = 0.5$, $w = d_{eff} = 5\mu$m the result is $l_c = 0.0215$m. At a twistpitch of 0.0254 m the loss factor $(l/l_c)^2$ would amount to 8.7%.

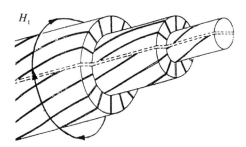

FIG. 3.7. Composite with twisted filaments fully coupled by H_t.

Filament twisting provides no help against **self-field coupling** as shown in Fig. 3.7; the filaments are fully coupled with respect to H_t. To decouple them would require one to transpose the filaments radially within the composite, which is excluded for practical reasons and manufacturing costs. Fortunately, transposition can be implemented at a later stage during the cabling process. The composites or wires are then transposed with a **transposition pitch** of $\sim 0.1\text{m} < 4L_t < 0.15\text{m}$, such that the net enclosed magnetic flux along $2L_t$ is close to zero. The principle has been adopted from classical windings for ac machines using Roebbel-type conductors and Litz-type wires.

In twisted composites an axial magnetic induction B_z is induced by the spiral-like wound filaments, allowing that a fraction of I_t flows in the inner, non-saturated part of the superconductor at an average j_{cz} [6]:

$$j_{cz} = \frac{I_t \pi}{4l^2}. \tag{3.23}$$

In modern Cu- (or Al-) stabilized NbTi and Nb$_3$Sn composites the self-field effect is masked, neutralized by the high thermal and electric conductivity of the matrix [10]. However, when the first NbTi composites with a high resistivity Cu-Ni matrix were developed 15–20 years ago for fast pulsed accelerator magnets with risetimes of a few seconds, premature quenching at low fields and considerable training were observed and identified as self field instabilities [8]. When designing NbTi composites for superconductive coils with rapidly changing magnetic fields, such as in the poloidal coils of 'Tokamak' fusion reactor coils, three component composites with high resistivity Cu-Ni and Cu matrices will be required. Nb$_3$Sn composites will in a similar way contain combined high resistivity (bronze) and Cu matrices and barriers.

4

MAGNETIZATION AND AC LOSSES IN SUPERCONDUCTORS

4.1 General remarks

Magnetization losses in superconducting filaments and other losses in superconductive wires, composites and cables in time-varying magnetic fields are of considerable importance to magnet design. Important theoretical and experimental investigations in this domain had been performed since the understanding of hard Type II superconductors due to the work of Bean and Kim (see chapter 2), Anderson *et al.* [1], which led to the development of the two main technical superconductors and and of other potential candidates, to be treated in Chapter 5.

In the early 1970s high energy particle accelerators with superconductive dipole and quadrupole magnets and with duty cycles close to those of accelerators equipped with classical magnets were investigated; rise times of ~ 1 s and duty cycles of a few Hz had been assumed. Loss calculations have shown that such 'fast' cycled machines would require more elaborate composites and cables and higher performance cooling systems compared to quasi-dc operated superconducting accelerators. It has been a fortunate coincidence that a parallel evolution pointed towards such an operation. Slow, long duration beam extraction and spilling techniques had been developed leading to cycles with longer flat tops of several minutes. When the advantages of particle colliders over fixed target machines became more and more evident [2,3,4] with flat tops extending to several hours or days, this mode of operation was also beneficial for the development of adequate superconducting dc magnets.

Rather stringent requirements are, however, imposed on specific excitation windings for rapidly varying magnetic fields in future nuclear fusion machines of the Tokamak type. Such magnet designs will not be treated in the present book.

Nevertheless, when designing superconducting magnets for accelerators and colliders or superconductive solenoids for large particle detectors, time varying magnetic fields and the related losses must be carefully considered during excitation and deexcitation.

4.2 Magnetization losses in superconducting filaments

One of the main loss components in a superconductive filament, when subjected to a time varying magnetic field, are the magnetization or hysteresis losses. One usually defines the magnetization M as the difference between the applied external field H_e and the average field H_i inside the superconductor:

$$M = \mu_0(H_e - H_i). \tag{4.1}$$

M is thus the magnetic induction due to the magnetization current $\pm I_c$ in the superconductor, flowing at critical current density j_c. A more accurate definition describes M as
the sum of the magnetic moments m per unit volume of the superconductor:

$$M = \sum \frac{i_{sc}\mu_0 s}{v} \tag{4.2}$$

with $s[\text{m}^2]$ the cross-section of an elementary superconductive loop carrying the super-
current i_{sc}.

In Type I superconductors with total field expulsion and supercurrents flowing only
within the penetration depth ξ (see Chapter 2) M is equal and opposite to the external
magnetic induction B_e:

$$M = \Lambda j_c \mu_0 = -B_e \tag{4.3}$$

as shown for a superconductive slab in Fig. 4.1. Such magnetization conditions are of
marginal importance in high field magnets as they only occur in very limited volumes of
excitation coils where B_e is close to zero. In Type II superconductors the magnetization
M can be calculated from the unit volume losses $w[\text{Wsm}^{-3}]$ represented by the surface
enclosed by the magnetization curve $M = f(\mu_0 H_e)$ [6,7]

$$\oint M(H_e)\mathrm{d}H = \oint H_e \mathrm{d}M. \tag{4.4}$$

W can also be computed from the electric field vector $\vec{E}$ and $\vec{j_c}$ as:

$$W = \int_0^t \left[\oint \vec{E}\vec{j_c}\,\mathrm{d}v \right] \mathrm{d}t. \tag{4.5}$$

For an assumed constant magnetization M and a triangular or trapezoidal shape of
the magnetic induction 0 - B_{em} - 0 as shown in Fig. 4.2 one obtains from eqn (4.4)

$$w = \frac{2M B_{em}}{\mu_0}. \tag{4.6}$$

As j_c varies considerably in that interval one should use eqn (4.5) with the average
current density $j_{c\,av}$ according to the Kim–Anderson relation:

$$j_{c\,av} = \frac{1}{B_{em}} \int_0^{B_{em}} \frac{j_{c0}B_0}{B + B_0} \mathrm{d}B = \frac{j_{c0}B_0}{B_{em}} \ln\frac{B_0 + B_{em}}{B_0} \tag{4.7}$$

with B_0 and j_{c0} obtained from the short sample characteristic of the superconductor.

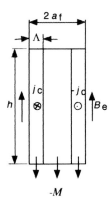

FIG. 4.1. Magnetization of a Type I superconductive slab.

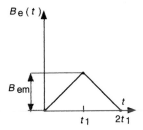

FIG. 4.2. Triangular magnetic induction pulse used for loss calculations.

Morgan [8] has calculated the energy loss w and the magnetization M for a super-conductive slab and a round wire of width diameter $2a_f$ placed in a time varying external magnetic induction as shown in Figs 4.3. For a **slab** of height $2b$ and length l one obtains from Maxwell's equation:

$$\mathrm{rot}\ \vec{E} = -\frac{\delta\ \vec{E_z}}{\delta x} = -\mu_0\frac{\delta\ \vec{H_e}}{\delta t}; \quad E_z = \mu_0\frac{\delta H_e}{\delta t}x \tag{4.8}$$

and for the power $P[\mathrm{W}]$

$$P(t) = 2bl \int_0^{a_f} j_c\mu_0\frac{\delta H_e}{\delta t}x\,\mathrm{d}x = \frac{a_f}{2}v\mu_0 j_c\frac{\delta H_e}{\delta t} \tag{4.9}$$

For a triangular field cycle according to Fig. 4.2 one obtains:

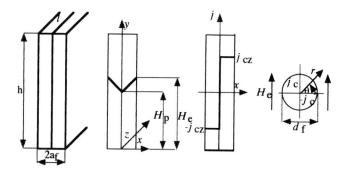

FIG. 4.3. Calculating magnetization losses in slabs and round filaments of Type II superconductors.

$$w = a_f \mu_0 H_{em} j_c; \quad M = M_0 = \frac{a_f}{2} \mu_0 j_c \qquad (4.10)$$

where from now on we define the magnetization in the superconductor by M_0. One can also define the field of full penetration H_p into the specimen:

$$H_p = a_f j_c \qquad (4.11)$$

and the ratio

$$\beta = \frac{H_p}{H_{em}}. \qquad (4.12)$$

For a **round** filament of diameter $2a_f$ according to Fig. 4.3b one obtains from eqn (4.8) expressed in cylindrical coordinates:

$$\frac{\delta E_z}{\delta r} = \mu_0 \frac{\delta H_{e\Theta}}{\delta t}; \quad E_z = \frac{\delta H_e}{\delta t} \mu_0 r \cos \Theta. \qquad (4.13)$$

The power loss is:

$$P = \frac{4}{3} j_c \mu_0 l a_f^3 \frac{\delta H_e}{\delta t} \qquad (4.14)$$

and the unit volume energy loss for the cycle $0 - B_{em} - 0$ is:

$$w = \frac{4}{3\pi} j_c B_{em} d_{eff} = \frac{2M_0 B_{em}}{\mu_0}. \tag{4.15}$$

The magnetization of a round superconductor is thus

$$M_0 = \frac{2}{3\pi} \mu_0 d_f j_c. \tag{4.16}$$

Wilson [6] has calculated the hysteresis loss in a round superconducting wire of diameter d_f placed in a magnetic field H_e parallel to the wire axis. For $H > H_{ep}$ (a parallel field of full penetration) the energy loss on a $0 - B_{em} - 0$ cycle amounts to

$$w = \frac{B_{em} j_{c\Theta} d_f}{3} \tag{4.17}$$

with $j_{c\Theta}$ the circumferential component of j_c. When the superconductor also carries a transport current I_t such that $I = I_t/I_c$, the unit volume hysteresis losses given by eqns (4.10, 4.15) and (4.17) are increased by the factor:

$$w' = w(1 + i^2). \tag{4.18}$$

4.3   Resistive losses in superconducting wires and composites

As already stated in Chapter 3 technical superconductors must be stabilized by incorporating into wires and composites a matrix of high electric and thermal conductivity such as Cu or Al. In time-varying magnetic fields eddy currents will be induced in these resistive matrices. In contrast to the hysteresis losses in the superconductor which only depend on the maximum magnetic induction excursion B_{em}, the eddy current losses are rate or frequency dependent. Let us first consider an axially symmetric composite. The filaments are embedded in a high conductivity matrix of resistivity ρ_m; the interfilamentary distance is w_f, and the fraction of the superconductor in the composite λ. This model is valid for most cases of NbTi filaments. In the case of Nb$_3$Sn filaments the fractions of the low resistivity matrix λ_{Cu} and of other components like bronze, Ta or Nb barriers have to be considered. For a round composite the equivalent matrix resistivity is:

$$\rho_e = \frac{\pi w_f}{3d_f} \rho_m. \tag{4.19}$$

For the same geometry, but with an outer (Cu-) ring of thickness Δr and ρ_m as shown in Fig. 4.4:

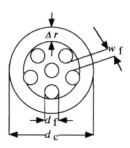

FIG. 4.4. Composite model for calculating ρ_e.

$$\frac{1}{\rho_{eq}} = \frac{1}{\rho_e} + \frac{\Delta r}{a\rho_m} + \frac{a\Delta r}{\rho_m}\left(\frac{2\pi}{l}\right)^2 \qquad (4.20)$$

with $4l$ the twistpitch length. Carr [9] has calculated the transverse resistivity ρ_t for two extreme cases of no contact resistance between matrix and filaments and for a fairly high contact resistance at the interface. The limiting values are:

$$\rho_{t1} = \rho_m\frac{1-\lambda}{1+\lambda}; \quad \rho_{t2} = \rho_m\frac{1+\lambda}{1-\lambda}. \qquad (4.21)$$

Wilson [10] has calculated the equivalent magnetization M_r due to eddy currents in a composite:

$$M_r = \frac{2\mu_0 l^2}{3\pi\rho_{eq}}\dot{B}_e \qquad (4.22)$$

and the losses w_r for a triangular cycle $0 - B_{em} - 0$:

$$w_r = \frac{4\mu_0 l^2}{3\pi\rho_{eq}}\dot{B}_e B_{em}. \qquad (4.23)$$

Wilson [6] has also calculated the losses in a composite exposed to an external longitudinal induction $\dot{B}_e$ and the losses due to the self-field $\dot{B}_t$. Untwisted or twisted filaments in a composite are not coupled by $\dot{B}_e$; they are, however, fully coupled by $\dot{B}_t$. In the first case the screening supercurrents will penetrate the composite flowing in the outside filaments and return through the inner ones—see Fig. 4.5—crossing the resistive matrix at the composite ends. It also shows the profile in the composite and the B_0 and j_c components. Since the time constant for this process

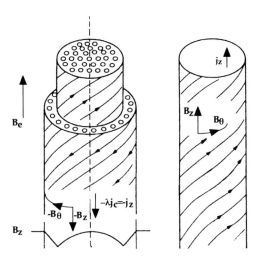

FIG. 4.5. Composite exposed to an external longitudinal induction $\dot{B}_e$.

$$\tau = \frac{\mu_0 l_w^2}{\pi^2 \rho} \qquad (4.24)$$

is of the order of 10^8 s, the resistive effect can be neglected and the losses will be hysteretic, independent of $\dot{B}_e$. For a field cycle $0 - B_{em} - 0$ the unit volume losses at full field penetration into the composite will amount to

$$w = \frac{B_{em}^2}{2\mu_0} \frac{d_c^2 \pi^2}{l^2} \left(\frac{1}{2\beta} - \frac{5}{12\beta^2} \right) \qquad (4.25)$$

with $4l$ the twistpitch length, d_c the composite diameter and

$$\beta = \frac{2\pi H_{em}}{\lambda j_c l} \geq 1. \qquad (4.26)$$

Equation (4.25) suggests that the longitudinal field losses could be reduced by increasing the twistpitch. This would, however, increase the transverse field losses which are predominant in high field accelerator magnets and in solenoids.

For the self-field or transport current losses one obtains for a cycle $0 - B_{em} - 0$ and full penetration of the composite:

$$w_t = \frac{B_m^2}{2\mu_0} \left\{ \frac{4}{i} - 1 + \frac{4(2-i)}{i^2} \ln \left[\frac{(2-i)}{2} \right] \right\}. \tag{4.27}$$

Self-field losses in composites are considerably smaller than transverse field losses due to the much higher values of the external field H_e compared to H_t. Large self field losses can, however, occur in cables made of non-transposed composites due to a remaining enclosed net self-field flux Φ_t. Fully transposed cables with a transposition pitch length $4L$[m] of the order of 10–15 cm are therefore essential. The composites will then interchange position with almost no Φ_t enclosed.

More calculations on magnetization and eddy current type losses in time varying fields have been made by Wilson [6]; cases of partial and of full filament penetration in round and slab type superconductors with and without superimposed transport currents are presented for a linearly changing external inductions B_e and in oscillating ones of the type $B_{em} \exp(i\omega t)$.

However, most superconducting high field magnets for accelerators and colliders are nowadays wound with flat cables, made of multifilamentary composites or strands; their magnetization and eddy current type losses can be calculated and measured as shown in the next section.

4.4   AC losses in flat superconducting cables

The **interstrand** eddy current losses in cables will now be considered and the concordant results of computations by G.Morgan [8], M.Wilson [10] and L.Krempaski [11] will be given. One considers a flat, rectangular cable of width $2c$ and thickness $4a$, a being the radius of the composite or strand. The cable is shown in Figs 4.6a and 4.6b. With the **magnetic induction** $B = B_z$ **perpendicular** to the wide face of the cable, two induced currents I_1 and I_2 can be distinguished: I_1 is induced in two strands crossing each other in points A, B, C and D with resistive contacts in A and C. Current I_2 flows in two adjacent strands with resistive crossings in points E and F. The average transverse contact resistivity ρ_a is given by

$$\rho_a = \frac{r}{2a} \tag{4.28}$$

with $r[\Omega\text{m}^2]$ the contact resistance times the unit area in between the crossing strands. The magnetization M_1 due to I_1 flowing in the loop ABCD induced by the component of the magnetic induction $\dot{B}_z = \dot{B}$, perpendicular to $2c$ is:

$$M_1 = \frac{4}{15} \frac{\mu_0 \alpha^2 L^2}{\rho_a} \dot{B}_\perp \tag{4.29}$$

with

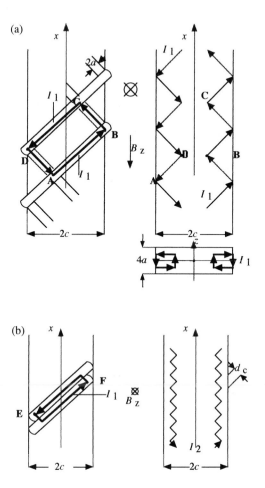

FIG. 4.6. Calculating the time dependent losses (a) in flat cables at $\dot{B}_e$; and (b) between adjacent strands in flat cables.

$$\alpha = \frac{c}{2a} \tag{4.30}$$

the aspect ratio of the cable, as shown in Fig. 4.7. For a triangular induction pulse $0 - B_{em} - 0$ the unit volume losses are

$$w_1 = \frac{8}{15} \frac{\alpha^2 L^2}{\rho_a} \dot{B}_\perp B_{em}. \tag{4.31}$$

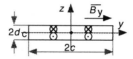

FIG. 4.7. Loss calculation at B_y parallel to the cable wide face.

The driving electric field E_z and the associated current density j_z have a maximum at the centre of the cable for $y = 0$, as shown in Fig. 4.6a.

The magnetization due to the current I_2 flowing in between adjacent strands as shown in Fig. 4.6b amounts to

$$M_2 = \frac{\mu_0 L^2 \dot{B}}{3 \rho_a cos^2 \theta} \approx \frac{\mu_0 L^2 \dot{B}_\perp}{3 \rho_a}. \tag{4.32}$$

and the unit volume losses are

$$w_2 = \frac{2}{3} \mu_0 L^2 \dot{B}_\perp B_{em}. \tag{4.33}$$

For a magnetic field H_e **parallel** to the cable wide face a longitudinal current I_3 will be induced in between the cable layers due to the flux variation Φ_e as shown in Fig. 4.7. The contact resistance ρ_a is the same as for M_2 since I_3 also circulates between adjacent strands. The magnetization M_3 due to I_3 is:

$$M_3 \approx \frac{\mu_0 L^2 \dot{B}_\parallel}{4 \alpha^2 \rho_a}. \tag{4.34}$$

The unit volume losses for a triangular pulse $0 - B_{em} - 0$ are

$$w_3 = \frac{1}{2} \frac{L^2 B_{em} \dot{B}_\parallel}{\alpha^2 \rho_a}. \tag{4.35}$$

The total magnetization M and the total losses w in a cable due to a time varying external field of arbitrary orientation with the components $\dot{B}_\perp$ and $\dot{B}_\parallel$—see Fig. 4. 8— can now be determined as:

$$M = \sqrt{M_\perp^2 + M_\parallel^2} \tag{4.36}$$

where

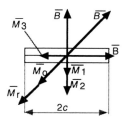

FIG. 4.8. Orientation of the B and M vectors.

$$M_\perp = M_0 \left\{ 1 + \frac{\dot{B}_\perp}{\lambda j_c d} \left[\frac{l^2}{\rho_e} + \frac{L^2}{\rho_a} \left(\frac{2\pi}{5} \alpha^2 + \frac{\pi}{2} \right) \right] \right\}. \qquad (4.37)$$

$$M_\parallel = M_0 \left[1 + \frac{\dot{B}_\parallel}{\lambda j_c d} \left(\frac{l^2}{\rho_e} + \frac{L^2}{\alpha^2 \rho_a} \frac{8}{3\pi} \right) \right]. \qquad (4.37b)$$

The losses for a triangular excitation cycle $0 - B_{em} - 0$ are:

$$\sum w = w_0 \left\{ 1 + \frac{1}{\lambda j_c d} \left[\frac{l^2 \dot{B}_{e\perp}}{\rho_e} + \dot{B}_\perp \frac{L^2 B_{em\perp}}{B_{em} \rho_a} \left(\frac{2\pi}{5} \alpha^2 + \frac{\pi}{2} \right) \right. \right.$$
$$\left. \left. + \frac{\dot{B}_\parallel B_{em\parallel}}{B_{em}} \frac{3\pi}{8} \frac{L^2}{\alpha^2 \rho_a} \right] \right\}. \qquad (4.38)$$

4.5 Experimental determination of the magnetization coefficients M and losses w in superconducting filaments, composites and cables

Having developed analytical expressions for the magnetization components M and the losses w in superconductive filaments, wires and cables in time varying magnetic fields, methods for their experimental determination will be given. Accurate knowledge of the various components of M and w should help in finding appropriate ways to reduce them. It is assumed that the filament, wire and cable geometry and composition, the orientation and time dependence of the external field and the $j_c = f(B_e)$ dependence are known. The parameters to be determined by adequate measurements are the filament magnetization M_0, the equivalent matrix and contact resistivities ρ_e and ρ_a, and the effective filament diameter d_{eff} which may be considerably larger than the filament diameter d_f notably in Nb_3Sn wires.

The magnetization can be measured by integration methods, small volume sampling and squid magnetometers; the losses during a magnetization cycle can be obtained by integrating the surface of the $M = f(B_e)$ loop or by calorimetric methods [7]. The principle of the integration method is shown in Fig. 4.9. Two identical pick-up coils are placed into a larger superconductive solenoid, providing B_e. A trimming coil provides

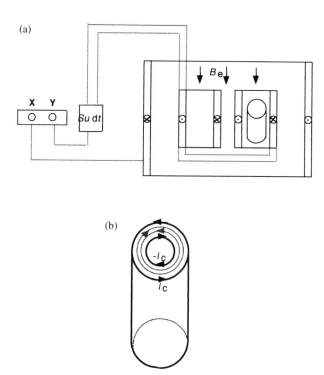

FIG. 4.9. (a) Setup for magnetization measurements in a thin superconductive cylinder; (b) flow $\pm I_c$ in the superconductive cylinder.

final balancing and no signal appears at the entrance of the integrator when varying B_e. The superconductive filament is wound on a cylinder of height h and placed into one of the pick-up coils. When B_e is turned on, the magnetization currents $\pm I_c$ will mutually cancel out except in the inner and outer layer. The sample can then be represented by a hollow cylinder with two azimuthal current sheets, as shown in Fig. 4.9b. The mutual inductances m_0 and m_i [H] between the sample and the pick-up coils can be determined by magnetic field computer programs. Any change in the external field ΔB_e will result in a ΔI_c in the sample and an unbalanced voltage signal u will appear at the integrator input:

$$\int u \, dt = \Delta I_c (m_0 - m_i) \tag{4.39}$$

yielding:

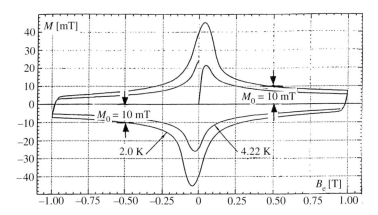

FIG. 4.10. Magnetization curve measured on a sample of a 0.5 mm Φ NbTi wire with $\lambda_{sc} = 0.385$.

$$\Delta M = \frac{\mu_0 \int u \, dt}{\lambda_{sc} h (m_0 - m_i)}.$$ (4.40)

λ_{sc} is the part of the superconductor in the sample. Measurements are performed at linear $\dot{B}_e$ cycles $0-B_{em}-0-(-B_{em})-0$ and curves shown in Fig. 4.10 are obtained. The curves refer to a 5 $\mu m \Phi$ NbTi wire with $j_c = 2 \times 10^{10}$ A m^{-2} at 0.5T and $\lambda_{sc} = 0.385$. Equation (4.16) yields $2M_0 = 0.0205$T, which agrees with the measured values. Due to the slow sweep of $\dot{B}_e = 0.01$T s^{-1} the eddy current losses can be neglected.

To separate the magnetization losses from the eddy current losses in the matrix given by eqns (4.15) and (4.23) one has to extrapolate the measured $2M = f(\dot{B}_e)$ values to $\dot{B}_e \to 0$ in order to obtain $2M_0$—see Fig. 4.11. The equivalent composite resistivity $\rho_e = f(\dot{B}_e)$ is obtained from the difference $2(M - M_0)$. The filament effective diameter can be obtained from the short sample curve of the superconductor $I_c = f(B_e)$, corrected by the amount of the self-field B_t.

$$B_t = \frac{\mu_0 \lambda I_c}{2\pi r}.$$ (4.41)

Added to B_e, one can determine d_{eff} as

$$d_{eff} = \frac{2M(B_e)s}{2\frac{2}{3}\pi I_c(B_e + B_t)}$$ (4.42)

with $s[m^2]$ the composite cross-section. The total magnetization losses per cycle are obtained by integration of the surface enclosed by the magnetization curve $M = f(B_e)$—see Fig. 4.10.

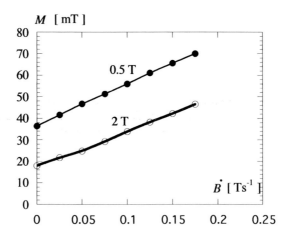

FIG. 4.11. Determination of $M_{\dot{f}=0}$ in the superconductor.

Extending the magnetization and eddy current loss measurements to **superconducting cables** made of wires or composites of known M_0, M_r, w_0, w_r, ρ_e and d_{eff} values, the average contact resistance of the cable can be found from eqns (4.35)–(4.38).

The losses in superconducting cables for high field magnets are an important part of the total heat losses, relevant for the design of the magnet cooling system. As an example for the limitation of these losses let us mention the planned next generation of superconducting fusion machines ITER. For a swing of the magnetic induction from 3T to -3T in 2 s, the magnetization and eddy current losses should be below 600 mJ cm^{-3}.

To conclude this chapter let us mention that **keystoned** cables are currently used in high field magnets for accelerators and colliders. As the keystoning angles are small— around and below 1°—the losses are calculated by assuming a flat cable of average thickness 2 d_{av}.

5

MANUFACTURING AND APPLICATION OF ADVANCED TECHNICAL SUPERCONDUCTORS

5.1 Introduction

Among the hard Type II superconductors described in Chapters 2 and 3 only fine filamentary NbTi and Nb$_3$Sn conductors are in worldwide use today as **technical superconductors**, where NbTi is used in 80–90% of all superconducting devices. Their manufacturing techniques and applications shall be presented with sufficient detail to help the reader in choosing a suitable superconductor for a specific application. Other cryogenic or low temperature superconductors which seem promising for future applications will be mentioned at the end of the chapter. High temperature superconductors—HTS—will not be extensively considered, notably because of their low overall current densities and the difficulty of manufacturing them in the form of long wires. However, their evolution should not be excluded, as some of them may in the future exhibit interesting properties at liquid helium temperatures. Their application to high field magnets in the more or less distant future should not be prejudiced or excluded.

The two technical supercoductors NbTi and Nb$_3$Sn do, in fact, meet the requirements for wide application in science, research and industry such as:

(i) the possibility of manufacturing 1–100 μm diameter single and multifilamentary composites, compact and hollow profiles and cables;

(ii) industrial manufacturing of large quantities or of long lengths of the above superconductors in a reproducible way within tight electrical, geometrical and mechanical tolerances with reliable verification and measuring facilities according to agreed upon criteria at any production stage;

(iii) reasonable manufacturing costs, notably for large quantity production; this is already the case for NbTi conductors and will hopefully soon be extended to Nb$_3$Sn when used in sufficient quantites.

5.2 Manufacturing of NbTi superconductors

Considerable convergence exists nowadays between the 'classical' manufacturing processes for fine filamentary NbTi wires, developed by different firms. This is notably true for the manufacture of high j_c wires with filaments below 10 μm - Φ, embedded in a high purity OFCD-copper stabilizing matrix of RRR values between 100 and 300. RRR is the ratio of copper resistivity at 300 K and 4.2 K. The superconductor is predominantly a high homogeneity Nb-46.5 to 47 weight % Ti alloy which is cheaper and

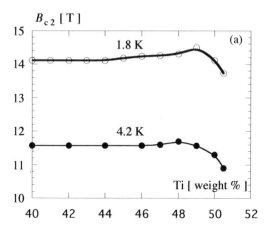

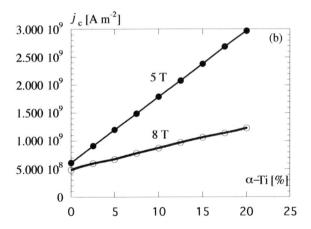

FIG. 5.1. (a) Influence of the Ti weight % on B_{c2}; (b) peak j_c-values depending upon
α-Ti[%].

exhibits better metalurgical properties than the 48-50 weight % Ti alloy of higher B_{c2}
values, as shown in Fig. 5.1 [1, 2].

Fine filamentary wires are obtained by single and double stacking processes, including cold working, conventional pressing and extrusion, annealing or heat treatment
stages, and twisting once the wire has been drawn to its final size. The starting element
is a **billet** of some 0.17–0.25 m diameter and 0.4–1.0 m length or height, yielding typically 100–250 kg of wire. The first or zero stage billet contains cylindrical NbTi rods,
placed into a predominantly copper matrix, whereby the rods may be protected by a
thin Nb barrier. The billets must be carefully assembled under vacuum and their copper

caps electron-beam welded. To obtain good bonding between the NbTi rods and the matrix and avoid 'sausaging' of the rods and voids, the billets are isostatically compacted, where this operation may be incorporated into the cold or hydrostatic process [3]. To obtain high j_c and B_{c2} values intermediate and final annealing steps at temperatures between 300 and 500°C are necessary for the precipitation of the normal conducting α-Ti isles, forming the required pinning centres. The temperatures and the number of annealing cycles should be as low as possible, as the formation of α-Ti degrades the mechanical properties of the superconductor and leads to the formation of intermetallic $Cu\text{-}Ti_2$ layers which in return reduce j_c [4, 5]. A Nb barrier will efficiently protect the superconductor against the formation of $Cu\,Ti_2$. D.C. Larbalestier *et al.* [4] have shown that three relatively short preannealing periods of 3 h, followed by a final annealing step of 480°C for 80 h, will result in the required, at least 20 volume % of α-Ti, proportional to j_c as shown in Fig. 5.1b. It has also been established that heat treatments are efficiently applied when the 'true strain' ϵ satisfies

$$\epsilon = \frac{\ln A_0}{\ln A} \leq 6 \qquad (5.1)$$

with A_0 [m^2] and A the initial and final wire cross-sections, obtained by cold or conventional pressing and extrusion [2, 3, 4, 5].

NbTi wires are manufactured by a single or double stacking process of hexagonal bars, preceded by the already mentioned zero stacking stage. First 10–100 cylindrical NbTi rods, placed into a Cu matrix are drawn to the size of the first hexagonal bar. A new billet is then formed with these bars, and the sides matched with a 1–2 mm tolerance [6]. This billet is then either drawn to the final wire size, heat treated and twisted, or shaped into a second hexagon for a second stage, final billet. Although wires with $\sim 20\,000$ filaments had been made by single stacking, a practical limit of $\sim 10\,000$ filaments seems more appropriate. The double stacking process is optimal for composites containing several $10\,000$ NbTi filaments of 2–5 μm diameter. In both cases hydrostatic extrusion or conventional indirect extrusion can be applied prior to final annealing. Figure 5.2 shows the manufacturing process sequence for NbTi wires. Figure 5.3 shows the cross-section of a 1.29 mm Φ NbTi wire for the LHC dipole magnets of CERN. The wire has 28 314 filaments of 5 μm Φ; the Cu:SC ratio is 1.8:1, the critical current density at 8 T and 4.2 K or at 11 T, 1.8 K amounts to $j_c = 1.114 \times 10^9$ Am^{-2} at a resistivity of $\rho = 10^{-14}\Omega$ m. Figure 5.4 shows a keystoned cable of 17 mm width and of 1.3/1.67 mm thickness with 40 such wires.

The **quality factor** n related to the superconductive to normal state transition is given by

$$n = \frac{\ln \rho}{\ln j_c}. \qquad (5.2)$$

For the above mentioned strands a quality factor of $n = 40$–50 has been measured on filament diameters of 5 μm $\leq d_f \leq 10\,\mu$m and of $20 \leq n \leq 30$ for 2.5 μm $\leq d_f \leq$

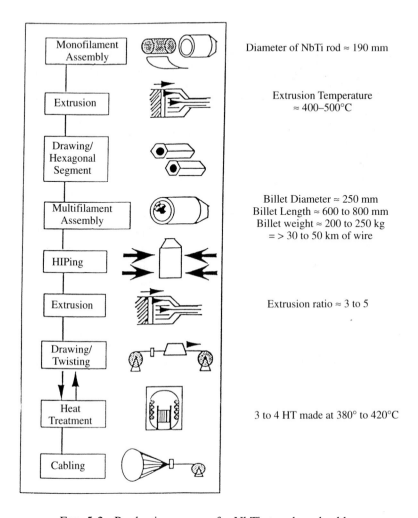

Diameter of NbTi rod ≈ 190 mm

Extrusion Temperature
≈ 400–500°C

Billet Diameter ≈ 250 mm
Billet Length ≈ 600 to 800 mm
Billet weight ≈ 200 to 250 kg
= > 30 to 50 km of wire

Extrusion ratio ≈ 3 to 5

3 to 4 HT made at 380° to 420°C

FIG. 5.2. Production process for NbTi strands and cables.

$3\,\mu m\Phi$. The j_c degradation due to conductor twisting has been measured to 3–5%. Cabling and cable compaction further reduced j_c by 5%, compared to j_c measured in non-twisted strands and corrected for the self-field or transport current effect.

Let us briefly mention the second group of fine filamentary NbTi conductors with mixed matrices for fusion application and superconductive devices exposed to fast varying magnetic fields [7]. Here conductors with overall time constants of $\tau \leq 1$ ms are required, which can be obtained by incorporating high resistivity CuNi barriers into the matrix. Plate 1 shows a mixed matrix conductor with a Cu:CuNi:NbTi ratio of 5:3:1.

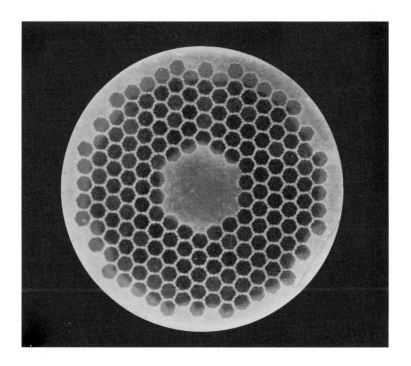

FIG. 5.3. Cross section of a Cu stabilized, 1.29 mm Φ fine filamentary NbTi wire with 28314 filaments of 5 mm Φ (courtesy: GEC–Alsthom Co., France).

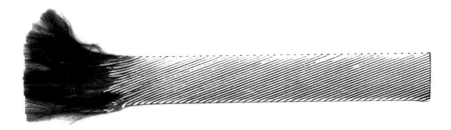

FIG. 5.4. A $1.3/1.67 \times 17$ mm^2 keystoned cable of 40 1.29 mm Φ strands. (Courtesy: GEC–Alsthom Co.)

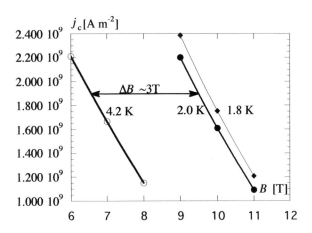

FIG. 5.5. $j_c = f(B)$ dependence of a 1.125 mm Φ fine filamentary NbTi wire at 4.2 K, 2 K and 1.8 K.

The respective time constant is 0.14 ms. The filament diameter amounts to $10\,\mu\mathrm{m}\Phi$. Conductors of this type are still in the development phase; their importance and complexity will increase with progress in fusion reactor technology.

Figure 5.5 shows the short sample lines $j_c = f(B_e)$ for a 1.125 mm Φ wire with $5\,\mu\mathrm{m}\Phi$ filaments measured at temperatures of $T_0 = 4.2$, 2.0 and 1.8 K. At the 2 K superfluid helium temperature one observes the characteristic $\Delta B = 3$T gain in magnetic induction at the same j_c:

$$[j_c(B_e)]_{T_0=4.2\,\mathrm{K}} = [j_{c4.2\,\mathrm{K}}(B_e + 3\mathrm{T})]_{T_0=2\,\mathrm{K}} \, . \tag{5.3}$$

The above short sample lines were measured in high quality, industrially produced filamentary NbTi composites with overall current densities of $2.75 \times 10^9 \le j_c \le 3 \times 10^9$A m^{-2} at 5 T and 4.2 K. Figure 5.6 shows the general three-dimensional j_c–T–B diagram for NbTi.

To extend the range of NbTi applications to higher magnetic inductions, **ternary** NbTi–Ta and NbTi–Hf alloys had been developed, processed and measured [1]. In fact, the performances of NbTi at 1.8 K and of Nb$_3$Sn at 4.2 K are comparable in the 8–11 T range, as shown in Fig. 5.7. The increase in magnetic induction expected for ternary NbTi was rather modest. In view of the more complex and expensive manufacturing costs the application of ternary NbTi alloys in this field range does not seem to be justified.

5.3 Artificial pinning centre-APC-NbTi superconductors

Good results have recently been obtained at the Lawrence Berkeley Laboratory (LBL) in association with a number of US superconductor manufacturers in the development of

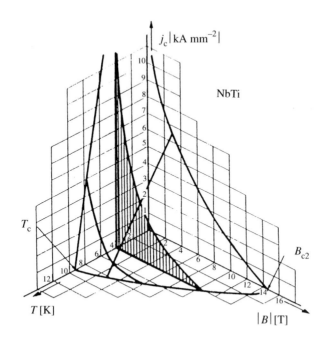

FIG. 5.6. Three-dimentional $j_c - T_c - B_c$ diagram of NbTi.

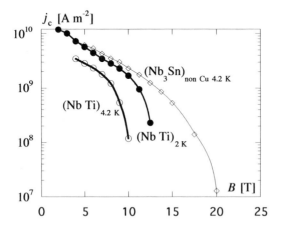

FIG. 5.7. Comparison of j_c figures of NbTi and of Nb_3Sn.

NbTi conductors made according to the APC approach [8]. Compared to conventional fabrication methods several advantages are expected such as: higher j_c values, simpler optimization of conductor performance, and lower production costs.

The idea behind the APC approach is to replace the α-Ti precipitation pinning centres by 'artifical' ones, which would be simpler to modify and optimize by choosing optimal distances and diameters of the pinning centres for a given magnetic field. Recent laboratory experiments have, in fact, shown that with well aligned and optimized pinning centres j_c values of up to 5×10^9 [A m^{-2}] could be obtained at 5 T and 4.2 K. Among several proposals the LBL laboratory retained a novel approach proposed by the 'Supercon' Company: the APC NbTi conductor is made of alternating sheets of Nb and Ti, rolled into a large monofilament, placed into an evacuated and electron-welded billet. The billet is cold or warm extruded and drawn to a hexagon, which is in turn cut, reassembled into a billet, and extruded and drawn to the final wire size **without any intermediate annealing**. Only one final annealing process is applied to the wire. The first APC method made wires exhibited promising properties, such as $j_c \approx 3.4 \times 10^9$ [A m^{-2}] at 5.6 T and 4.2 K and 1.62×10^9 [A m^{-2}] at 7 T and 4.2 K. APC NbTi wires of 1.14 mm Φ and cables made of 0.91 mm and 0.65 mm Φ wires have been tested in a small solenoid and a 1 m long, 5 cm aperture dipole magnet. The 8 T solenoid, wound with the 1.14 mm Φ wire, performed as expected [9]. The overall performance of the two cables, used in the 1 m dipole magnet was $\sim$ 10% lower compared to a previous identical magnet wound with 'conventional' NbTi cables. Analysis has shown that the reduced performance of the APC method processed cables was due to insufficient billet compaction, too large filament spacing, and sausaging of not enough uniform filaments. Another important issue is the production costs. A detailed comparison has shown that APC processed NbTi wires should be 50% cheaper than conventionally produced wires of the same performance [8, 9].

5.4   Intermetallic A15 compound superconductors Nb$_3$Sn and V$_3$Ga

The considerable potential of superconductors, based on intermetallic A15 compounds, was early recognized, notably their higher T_c and B_{c2} values compared to NbTi [10]. In fact, the first commercialy produced supercoductor by Kunzler *et al.* [11] almost 50 years ago was a Nb$_3$Sn wire, obtained by packing Nb and Sn powder into a Nb tube and drawing it to the final diameter. To overcome the drawback inherent in all A15 superconductors, their brittleness and absence of any ductility after reaction, the small coils of Kunzler were wound with a non-reacted conductor and than reacted or heat treated *in situ*.

The next step was the development of Nb$_3$Sn and to a lesser extent of V$_3$Ga tapes by chemical vapour deposition and diffusion of Sn into a Nb (or of Ga into a V) substrate, welded to or sandwiched between stabilizing Cu strips [12, 13, 14]. Due to the large width-to-thickness ratio, the tapes did not meet the stability criteria mentioned in Chapter 3 at magnetic inductions normal to the tape width. However, Nb$_3$Sn tape wound and dc-operated high field solenoids with the tape narrow sides matched to the magnetic field pattern, notably at the coil ends, had been designed and successfully operated. Two decades after the first powder made wires by Kunzler, fine filamentary, intrinsically stable Nb_3Sn composites were developed as well as coil winding techniques and materials which met the challenges and requirements related to the reaction process [15].

The development of fine filamentary V_3Ga composites was technically and commercially less successful. With the exception of a few inserts in high field test solenoids only Nb_3Sn fine filamentary wires and cables are nowadays used in superconducting magnets where fields beyond the capability of NbTi windings at 4.2 K and 1.8 K are required.

5.5 Manufacturing of fine filamentary Nb₃Sn composites

Before describing the manufacturing processes for Nb_3Sn composites, one has to clearly define the (critical) current densities in this superconductor. In the essentially two component, Cu stabilized NbTi superconductor one has to distinguish $j_{c\,NbTi}$ from the overall or average $j_{c\,av} = \lambda j_{c\,NbTi}$, with λ the volume fraction of the superconductor. Nb_3Sn composites have a third component, bronze. One has thus to disintinguish between $j_{c\,Nb_3Sn}$, roughly the same for a given B and not depending upon the manufacturing process, $j_{c\,non\,Cu}$ standing for the j_c in the Nb_3Sn plus bronze part, and finally the average $j_{c\,av}$ in the composite, including the stabilizer. An increase of $j_{c\,av}$ will mainly depend upon the ability to reduce the bronze content after reaction.

To this end different manufacturing processes had been developed; their choice will depend upon the required parameters of the composite, such as small effective filament diameters d_{eff}, high $j_{c\,av}$, and tolerable magnetization and ac losses. It will be shown that some of these requirements are contradictory and that a compomise will have to be found for each case of application. Figure 5.8 shows the general j_c–B_{c2}–T_c diagram for Nb_3Sn and NbTi.

5.5.1 *The bronze process*

The bronze route procedure for manufacturing fine filamentary Nb_3Sn composites is very similar to the fabrication of NbTi wires. Billets of ductile precursor Nb rods, assembled in a bronze matrix, are extruded and drawn to the final wire size in, usually, a two-stage process. However, the Sn content in ductile bronze is limited to less than 15%, normally to 13.5%, and a large bronze matrix to Nb ratio of 3:1 is required to provide sufficient Sn to the Nb rods. The bronze matrix can be incorporated in two ways, locally by choosing a ratio of $0.6 < \alpha < 1.5$ and by adding the remaining $3 - \alpha$ part on the periphery. At $\alpha \approx 0.6$ the d_{eff} of the filament will increase by about a factor of two with respect to d_f, due to filament bridging, related to irregular deformations during cold and hot working and notably because of the 38% volume increase during the reaction of Nb into Nb_3Sn. Choosing a higher ratio of $1.2 < \alpha < 1.5$ will reduce d_{eff} to values close to d_f [16, 17, 18]. The bronze route is usually a two-stage process; the initial billet with several hundred Nb rods is drawn to a hexagon of intermediate size which is then cut and the shortened elements are reassembled in a second billet to be

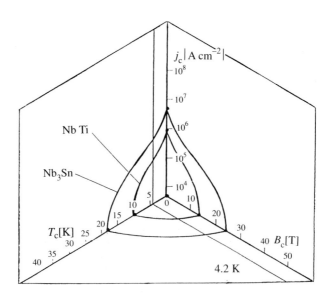

FIG. 5.8. The $j_c - T_c - B_c$ diagram of Nb_3Sn and Nb_3Ti.

extruded, annealed and drawn to the final size. Wires with up to 50 000 filaments had been drawn to diameters of 1.5–3.0 mm Φ.

Like NbTi composites, Nb_3Sn wires must be dynamically stabilized against flux-jumping, predominantly by OFCD copper. In special cases, where high transparency conductors are required, a high-purity Al stabilizer is used. The Cu stabilizer must be protected against diffusion of the bronze–Sn by a thin Ta or Nb barrier of a few μm. Nb barriers are more ductile and cheaper, but have the disadvantage of being supercon-ductive at low magnetic inductions, which may adversely influence the field precision in specific cases like NMR magnets and particle accelerator magnets. Ta barriers are therefore preferred. The stabilizing Cu may be incorporated **internally** with up to 27% Cu of the wire cross-section or **externally** with a Cu part of 30–60%.

During the reaction process when Sn diffuses into the Nb, two competing phenom-ena are taking place. An increasing layer of Nb_3Sn tends to increase j_c, whilst the growth of the grain size tends to reduce it, as the pinning centre density in the su-perconductor decreases. The reaction process can be optimized for a given B. Typical reaction temperatures are $650°C < T_r < 700°C$ during $t_r \approx 100$ h. The optimal reaction cycle is usually established by the wire manufacturer and communicated to the magnet designer.

Figure 5.9 shows a 0.7 mm Φ composite with 6156 filaments of 3–5 μmΦ, and an internal 23% Cu stabilization, protected by a Ta barrier. Figure 5.10 shows a 1.38 mm Φ wire with 50 000 such filaments and a 38% external Cu stabilization, again protected

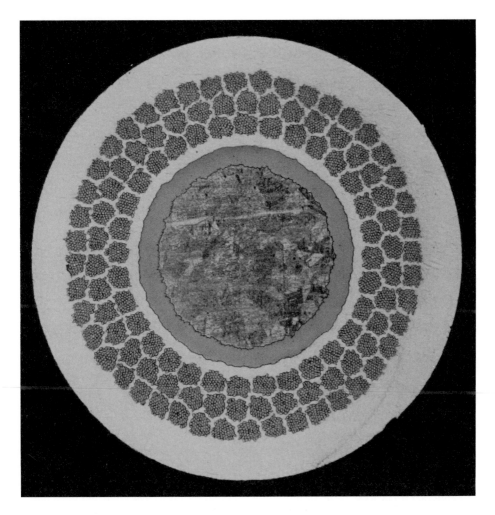

FIG. 5.9. A 6156 filament Nb₃Sn composite with internal Cu stabilization and Ta barrier (courtesy: Vacuumschmelze GmbH, Germany).

by a $\approx$ 4 μm Ta barrier and a cable made of 28 such strands.

For high field applications beyond 12 T the performance of bronze processed Nb₃Sn ckmposites can be improved by adding $\approx$ 1 weight % Ti or 7.5 weight % Ta to the Nb rods. The addition of Ta is prefered as it yields a more ductile ternary (Nb Ta)₃ Sn alloy. The $j_{c\,non\,Cu}$ figures are shown in Fig. 5.17a below.

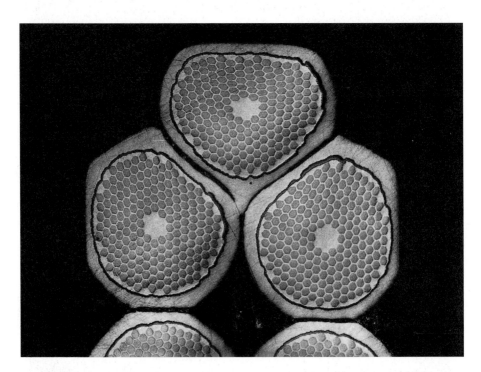

FIG. 5.10a. Cross-section of 1.38 mm Φ, 50 000 Nb$_3$Sn filament composites with external Cu stabilization forming part of a 28 strand cable. (Courtesy: Vacuum-schmelze GmbH, Germany).

5.5.2 *The internal Sn process*

The internal Sn process is the result of concentrated efforts to overcome the main constraint of the bronze route, the 13.5 weight % limit of the Sn content. The basic idea was to develop Nb$_3$Sn conductors from configurations with distributed local Sn sources surrounded by Nb rods and located in tubes of Cu or CuNi, as shown in Fig. 5.11. These tubes are drawn to hexagonal subelements. Several hundred shortened hexagons are re-assembled into a second billet, which contains an outer stabilizing Cu ring, protected by a suitable barrier. Since all components of the composite are ductile, they can be drawn to the final wire diameter **without intermediate annealing**. Figure 5.12 shows a 0.765 mm Φ wire with seven Sn sources, placed in a Cu plus Nb matrix, surrounded by a Ta barrier protecting the outer stabilizing Cu ring.

The performance of internal Sn-route made composites can be enhanced in several ways: adding a small amount of Cu, in the form of sheets, rolled around the Nb rods will increase $j_{c\,non\,Cu}$ at lower and medium fields [19, 20, 21].

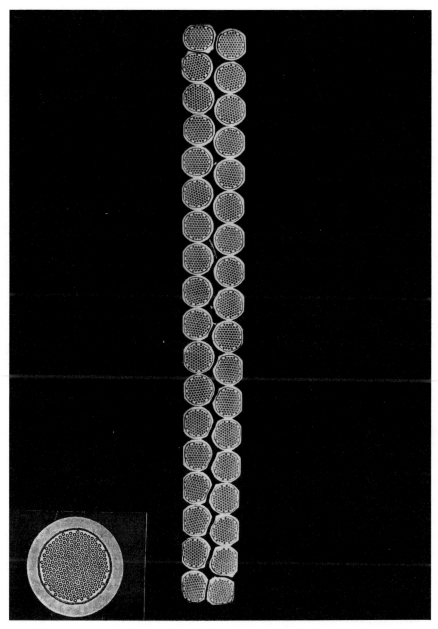

FIG. 5.10b. Cross-section of a Nb$_3$Sn cable made of 36 0.92 mm Φ strands with 20 000 filaments and external Cu stabilization. (Courtesy: Vacuumschmelze GmbH, Germany).

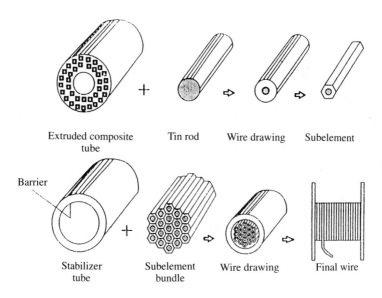

FIG. 5.11. Schematic presentation of the internal Sn process.

In the refined **tin tube source** or **TTS** variant [21] the outer part, consisting of the Cu stabilizer with an internal Nb foil wrap, and the inner part, containing the assembled hexagons of single Nb rods in Cu cans, are drawn separately. The inner part is then wrapped into an Sn foil, placed into the outer, stabilizing part and drawn to the final wire size, as shown in Fig. 5.13. Sn is thus provided from the internal sources and from the outer wrap. In the low and medium field range the conductor performance can be enhanced by adding Cu to the Nb rods and in the high field range by adding $\approx 1.3\%$ of Ti [21]. A comparison of the $j_{c\,non\,Cu} = f(B)$ values for Nb$_3$Sn composites made by the TTS procedure is summarized in Fig. 5.15.

5.5.3 *The modified jelly roll or MJR process*

The **MJR** process is an original variant of the internal Sn method for producing fine filamentary Nb$_3$Sn wires, pursued by the Teledyne-Wah Chang Company in the US. The principle is shown in Fig. 5.14. Two sheets of an expanded Nb mesh and of Cu are rolled in parallel around a solid Sn rod (or a high Sn content bronze rod). By so doing the Nb part in the non-Cu area can be increased to $\approx 35\%$. The complete roll inserted into a Cu tube forms the billet. During the drawing process the Nb diamond cross-section is reduced ≈ 750 times and elongated from 0.025 m to 1500 m. The billets are drawn to hexagons of $\approx 0.012\,m\Phi$, reassembled in a second billet which is by a multidrawing process reduced to the final diameter of 0.25 mm Φ–1.5 mm Φ. No annealing is required even at a 10^5 reduction of the initial cross-sectional area [22]. The initially pursued MJR process with a bronze core has been abandoned, as it required frequent

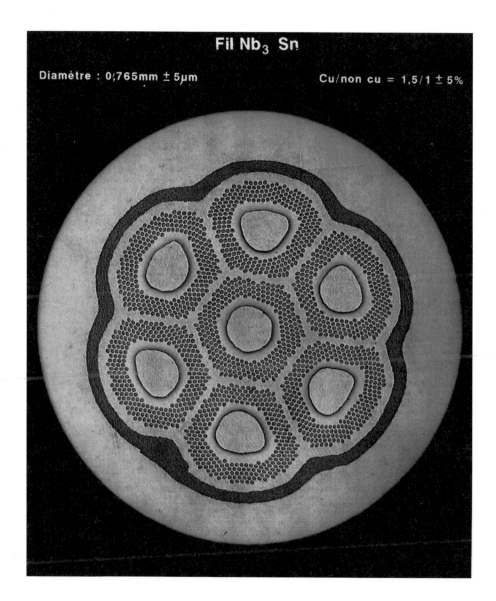

F$_{IG}$. 5.12. A 0.765 mm Φ Nb$_3$Sn filamentary composite made by the internal Sn process (courtesy: Alstrom–Intermagnetic Co., France.)

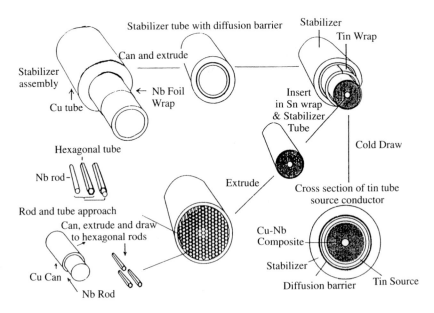

FIG. 5.13. Schematic presentation of the tin tube source (TTS) process for manufacturing filamentary Nb$_3$Sn composites.

annealing leading to premature reaction of Sn with Nb and to filament degradation. In addition, the large bronze-to-Nb volume significantly affected the $j_{c\,av}$ in the composite. Protecting barriers must be inserted between the Sn core plus the Nb and Cu sheets and the peripheric Cu stabilizer in order to protect it from Sn penetration and contamination. To this end metallic barriers of Ta, V and Nb are used.

By varying the parameters in the MJR processed Nb$_3$Sn composites like the density of the Sn rolls, the (increased) Sn fraction, and recently also by incorporating not entirely reacted Nb barriers, four wire qualities are available today [22]:

(i) low $j_{c\,non\,Cu}$, low hysteresis loss wires of $d_{eff} \approx 20\,\mu m\Phi$;

(ii) high $j_{c\,non\,Cu}$, high hysteresis loss wires with $60\,\mu m\Phi \le d_{eff} \le 100\,\mu m\Phi$;

(iii) increased Sn fraction, high $j_{c\,non\,Cu}$, high hysteresis loss wires with $70\,\mu m\Phi \le d_{eff} \le 110\,\mu m\Phi$;

(iv) increased Sn fraction, reduced Sn core element wires of high $j_{c\,non\,Cu}$ with hysteresis losses corresponding to filament diameters of $46\,\mu m\Phi \le d_{eff} \le 55\,\mu m\Phi$.

Recently high values of $j_{c\,non\,Cu} \approx 2.5 \times 10^9\,A\,m^{-2}$ have been measured in these wires at 10 T. This true breakthrough has been obtained by replacing the V barriers by Nb. The $j_{c\,non\,Cu}$ values measured in the bronze route, MJR, and various internal Sn-method processed Nb$_3$Sn composites are compared in Fig. 5.15. However, this comparison would

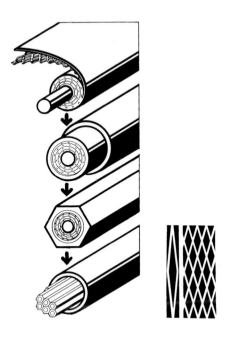

FIG. 5.14. Principle of the modified jelly roll (MJR) process (courtesy: Teledyne Wah–Chang Co., USA).

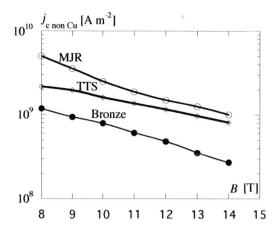

FIG. 5.15. Best $j_{c\,non\,Cu}$ values obtained by different manufacturing methods of Nb₃Sn wires.

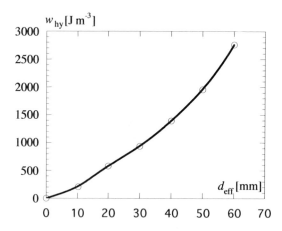

FIG. 5.16. Dependence of magnetization losses w_{hy} upon effective filament diameter d_{eff}.

be biased without comparing the hysteresis losses and the d_{eff} figures for the different wires, as shown in Fig. 5.16.

Confronted with these options, the magnet designer should be in the position to choose the appropriate Nb$_3$Sn composite and cable, meeting the magnet requirements. In the case of dc or quasi-dc operated high energy physics accelerator and collider magnets with long rise times one will opt for composites yielding the highest possible $j_{c\,non\,Cu}$ values, provided the related d_{eff} and hysteresis loss figures are acceptable and the field distortions due to the magnetization currents in the superconductor can be compensated by correcting coils. In fast cycled magnets for fusion devices limits will be set for the hysteresis and other dynamic losses in the composites, leading to the choice of Nb$_3$Sn wires of smaller d_{eff}, with lower losses but also of reduced $j_{c\,non\,Cu}$ values.

5.5.4 *The Nb powder method*

This method was developed about 15 years ago by the Netherland Research Foundation (ECN) at Petten and at the University of Twente [23]. The basic idea is very simple: hollow Nb tubes were filled with fine granulated Nb$_2$Sn powder and placed into a Cu matrix. Compared to previous methods, a great proportion of the non-active bronze content could be eliminated and replaced by more superconductive elements or by more stabilizing Cu. The non-reacted part of the outer Nb tube acts as a barrier between the Cu and the Nb$_3$Sn. Voids and hollow zones in the centre of the Nb$_3$Sn parts are inherent to this method.

Strands with diameters up to 1.3 mm Φ have been manufactured and very promising values of $j_{c\,av} \approx 10^9$ A m^{-2} have been obtained at 10 T and 4.2 K. Filament and wire diameters are closely related: 15 μm diameters were obtained in wires of 0.3 mm

Φ, 35–40 μm in 0.9 mm Φ wires, while in 1.25 mm Φ wires the filament diameter exceeded 50 μm. In the early cables damaged strands at the cable narrow edge had been observed. Once the bonding between the Cu matrix and the hexagonally located Nb$_3$Sn filaments had been improved by optimizing the heat treatment, reliable strands could be manufactured in lengths required for superconductive cables in high field magnets. Recently a 1 m long, single bore high field magnet wound with powder method processed strands reached a bore field of 11.1 T.

5.5.5 *Nb$_3$Sn superconductors at temperatures below 2 K and their application to very high fields*

Concluding this section, the operation of Nb$_3$Sn superconductors at superfluid helium temperatures below $T < 2\,\mathrm{K}$ is discussed. Compared to the 3 T gain for NbTi superconductors in the 10 T range when operated below 2 K, the gain in j_c or B for Nb$_3$Sn amounts to only 15–19%. Important progress has, however, recently been achieved with Ti-doped, and notably with Ta-doped (Nb 7.5 weight % Ta)$_3$ Sn strands cooled by superfluid helium in the $15\mathrm{T} \leq B \leq 25\mathrm{T}$ range [24, 25]. This development has been triggered by the need for high resolution spectrometer solenoids in the 20–25 T range for medical imaging. As shown in Figs 5.17a and 5.17b, Ta-doped Nb$_3$Sn strands, cooled below 2 K, exhibit a gain in $j_{c\,\mathrm{non\,Cu}}$ of more than a factor of two compared to cooling at 4.2 K. The values were measured on strands made by the internal Sn and TTS methods. It is therefore desirable to further investigate and improve the manufacturing of 'doped' Nb$_3$Sn wires made by the internal Sn, TTS, MJR or powder methods. This is essential for the development of the next generation of 15–20 T superconductive magnets to eventually increase the energy of the actual particle accelerators and colliders. If concentrated effort is still required for reliable manufacturing of high j_c strands and cables in that field range, it is encouraging that the $j_{c\,\mathrm{non\,Cu}}$ values recently measured at 15 T and 1.8 K correspond to figures obtained some years ago in bronze processed strands at 10 T and 4.2 K.

First results on artificial pinning–APC–microstructures of Ti- and Y-doped thin **Nb$_3$Sn** films indicated j_c values of $4 \times 10^9\,\mathrm{A\,m^{-2}}$.

5.5.6 *The critical current dependence upon transverse pressure in Nb$_3$Sn strands and cables*

It has already been stated that Nb$_3$Sn superconductors belong to the group of A15 materials, which become brittle after the heat treatment process. Due to the complex structure of Nb$_3$Sn wires containing different elements like bronze, copper and Ti or Nb barriers, experts have been concerned about the longitudinal stress and transverse pressure influence on the critical current in a specific wire. The influence of **longitudinal** stresses was first measured by Rupp *et al.* in 1977 [26]. If j_{c0} is the critical current density in a mechanically unstrained, reacted wire, the $\frac{j_c}{j_{c0}} = f(\sigma_1)$ or $\frac{j_c}{j_{c0}} = f\left(\frac{\Delta l}{l}\right)$ behaviour is shown in Fig. 5.18. Below a critical stress σ_{cr} or strain $\left(\frac{\Delta l}{l}\right)_{\mathrm{cr}}$, $\frac{j_c}{j_{c0}}$ increases to a maximum value $\frac{j_{cm}}{j_{c0}}$ and then decreases for $\sigma_1 > \sigma_{\mathrm{cr}}$. This behaviour can be explained by the

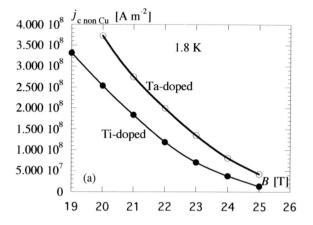

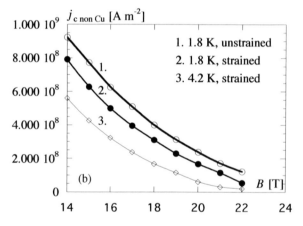

FIG. 5.17. (a) $j_{c\ nonCu}$ values, measured in Ta- and Ti- doped, TSS-processed Nb_3Sn strands at 1.8 K; (b) $j_{c\ non\ Cu}$ in Ta-doped, TSS-processed Nb_3Sn strands in the high field range at 4.2 K and 1.8 K.

pressure exercised by the surrounding bronze on a reacted Nb_3Sn filament due to their different thermal contraction coefficients. Pulling the wire in the longitudinal direction the internal pressure is first reduced and $\frac{j_c}{j_{c0}}$ is increased. Increasing σ_1 beyond σ_{1c} will reduce $\frac{j_c}{j_{c0}}$ in the Nb_3Sn filament and finally break it. In high field solenoids wound with Nb_3Sn wires and energized to the nominal field the longitudinal stress should remain below $\sigma_1 < \sigma_{1c}$ and the operating region should be on the left side of Fig. 5.18. In high field accelerator magnets wound with Nb_3Sn cables longitudinal stresses appear at the coil ends without influencing in a significant way the mechanical containment of the

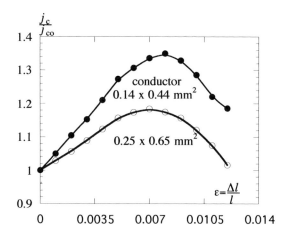

FIG. 5.18. Influence of the longitudinal strain ε on the critical current density j_c in Nb_3Sn superconductors.

end forces.

The application of a **transverse** pressure $-p_{tr}$ to a reacted Nb_3Sn strand can affect the critical current density $\frac{j_c}{j_{c0}}$ in a significant way, as it may lead to deformations of strands and filaments, damage the barriers, and short-circuit adjacent filaments. The effect has been measured by Specking and Ekin [27, 28]; a typical result for a 0.4 mm Φ wire is shown in Fig. 5.19. The results raised considerable concern about the use of Nb_3Sn wires for high field accelerator magnet windings, which are always exposed to high transverse pressures $-p_{tr}$. However, high field accelerator magnets are wound with insulated and epoxy resin impregnated Nb_3Sn cables where the transverse pressure is distributed over the entire cable surface. The influence of the transverse pressure on the critical current in Nb_3Sn cables $\frac{-\Delta I_c}{I_c} = f(-p_{tr})$ has been measured by several authors [29, 30, 31, 32] and the results are shown in Figs 5.19–5.20. The measurements have also shown that the effect is field-dependent: $\frac{-\Delta I_c}{I_c} = f(-p_{tr,B})$. The comparison of results obtained for different Nb_3Sn cables according to Figs 5.20 and 5.21 shows the influence of the wire manufacturing process and of the keystoning angle. Cables made of Nb_3Sn filaments which are well embedded and protected by a bronze or copper matrix, such as bronze and powder method processed wires, will in general exhibit a smaller transverse pressure effect compared to wires made according to various internal Sn processed strands. Cables with large keystoning angles exhibit a higher degradation of I_c than flat cables, made of identical strands.

5.6   Other high field superconductors

In this section a few more potential high field superconductors are mentioned, which had either been used in a limited number of experimental magnets or could be of interest, if

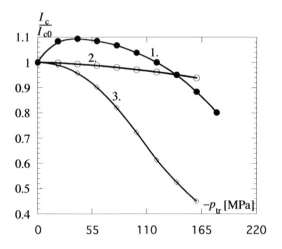

FIG. 5.19. Comparison of the $I_c = f(-p_{tr})$ depedence in epoxy-impregnated Nb_3Sn cables (1) and (2) and on a bare strand 3.

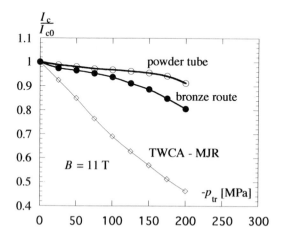

FIG. 5.20. Influence of Nb_3Sn manufacturing methods on the sensitivity of epoxy-impregnated cables to $-p_{tr}$.

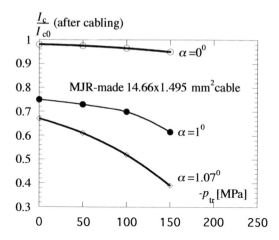

FIG. 5.21. Keystoning influence on the sensitivity of an Nb_3Sn impregnated cable to $-p_{tr}$.

further developed and improved.

Fine filamentary **Nb_3Al** or **$Nb_3(Al, Ge)$** superconductors of higher T_c and B_{c2} than Nb_3Sn can be produced in several ways. In the solid state and liquid quenching processes wires and ribbons are heated to 1200–1700°C, followed by a long annealing process at $\approx$ 700°C. In another variant a melted Nb 25% Al liquid is welded on to a substrate. $j_{c\,non\,Cu}$ values of 10^9 A m^{-2} have been measured at 20 T and 4.2 K. However, the liquid quenching process is incompatible with Cu stabilization and is thus unpractical.

In the jelly roll process a bilayer of 75 μm and 25 μm thick Nb and Al ribbons was rolled around a pure Cu rod to a diameter of $\approx$ 8 mm.

The rods were then bundled in a Cu can, drawn to hexagons, which were in turn assembled in a Cu billet and drawn to the final wire size—see Fig. 5.22. The $j_{c\,non\,Cu}$ tends to increase in wires of reduced diameters, but the drawing process then becomes increasingly difficult. The best obtained $j_{c\,non\,Cu}$ values amounted to 1.4 10^9 A m^{-2} at 8 T.

In the clad chip extrusion process shown in Fig. 5.23, a $\approx$ 1 mm Nb sheet is roll-bonded between two 0.14 mm Al sheets and then cut into chips, placed into a Nb loaded Cu can, double extruded, and drawn to wires of $\approx$ 3 10^6 times the reduced cross-section. A 1373°C, - 1 h; plus a 1000°C, 96 h, heat treatment is then applied to the wire. $j_{c\,non\,Cu}$ values of 10^8 A m^{-2} at 17 T, 4.2 K and of 10^9 A m^{-2} at 10 T, 4.2 K have been obtained. Similar results were obtained on powder metallurgy processed Nb_3Sn wires, where Nb and Al powders of 150 μmΦ and of 9 μmΦ were mixed together, placed into a Cu can, extruded, and heat treated at 750°C $\leq$ T $\leq$ 900°C.

Concluding this chapter two more high field superconductors are mentioned and

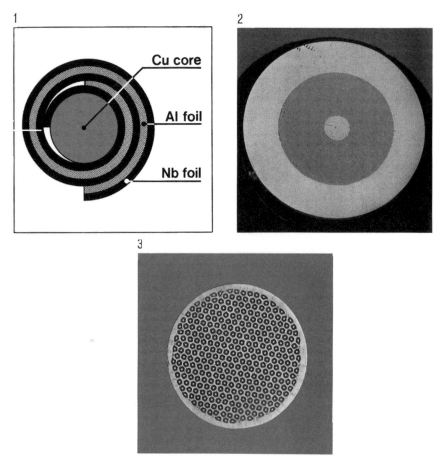

FIG. 5.22. Manufacturing process of fine filamentary Nb₃Al wires (courtesy: LMi Co., Italy).

their prospects briefly discussed: the **Chevrel phase** and the **oxide based supercon-ductors**, operated at liquid helium temperature.

Chevrel-phase superconductors of a general Pb_{1-x} Sn_x Mo_6 S_8 composition are obtained by elemental powder reaction at $\approx 800°C$ and a pressure of 2 kbar, followed by a post-annealing process at temperatures up to 300°C during ≈ 1000 h to promote grain growth [33, 34]. The main feature of Chevrel-phase superconductors is the high B_{c2} of 40 T $\leq B_{c2} \leq$ 50 T. Chevrel-phase superconductors are brittle and must be bonded to a substrate which reduces the overall $j_{c\,av}$. In the superconductive part j_c values of $0.8 \times 10^9 \leq 1.1 \times 10^9$ A m^{-2} have been measured at 9 T, 4.2 K and of $1.2 \times 10^9 \leq j_c \leq 1.5 \times 10^9$ A m^{-2} at 9 T, 2 K. With the necessary addition of a substrate and a stabilizer the actual performances of Chevrel-phase superconductors

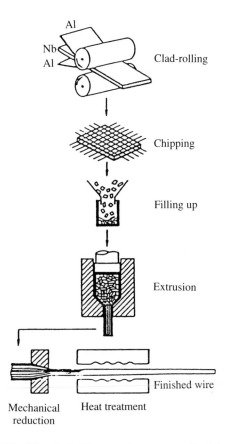

Al

Nb

Al

Clad-rolling

Chipping

Filling up

Extrusion

Mechanical
reduction

Heat treatment

Finished wire

FIG. 5.23. The clad-chip extusion process for Nb$_3$Al wires.

in the 10–14 T range are below those of fine filamentary Nb$_3$Sn wires. A number of technical, metallurgical and other problems have still to be solved. It is believed that an enhancement of j_c will depend upon improving the still poor intergrain connection, the optimization of the heat treatment at relatively low temperatures, the production of fine starting powders, and on the elimination of local defects in the μm range. [33, 34, 35, 36]. In addition, problems related to the industrial fabrication of wire lengths of several km with an adequate stabilizer and barriers and of the wire cabling will have to be solved. A good review on Chevrel-phase superconductors is given in [37].

Among high temperature (**HTS**) superconductors which look promising when operated at 4.2 K, hot or cold rolled and pressed silver sheathed Bi (2223) tapes are mentioned. They are being developed at the University of Geneva, Switzerland [38, 39, 40], by the American Superconductor Corporation (ASC) [41] and the National Laboratory

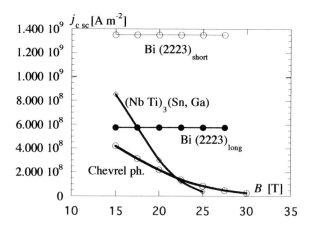

FIG. 5.24. $j_{c\,sc}$ values for Bi (2223) tapes, Chevrel phase, and (NbTi)$_3$ (Sn, Ga) superconductors at 4.2 K.

for High Energy Physics at Tsukuba, Japan [42]. Once improved, these superconductors
may be of interest for accelerator magnets and solenoids beyond 15 T. A comparison
of the $j_{c\,sc}$ values referring to the superconductive part obtained in Bi (2223) tapes, the
Chevrel-phase, and in Ti doped (NbTi)$_3$ (Sn or Ga) superconductors is given in Fig. 5.24
[40]. Considerable research and development is needed to solve the main problems of
4.2 K operated HTS conductors such as the $j_{c\,sc}$ dependence upon strain and field orientation and the still low overall $j_{c\,av}$ in the superconductor plus substrate cross-section.
The problem of reliable industrial manufacturing of wires and tapes in required lengths
has yet to be solved.

6

COOLING OF SUPERCONDUCTING HIGH FIELD MAGNETS

6.1 General remarks about losses in superconducting magnets and on the cooling modes

Efficient cooling of superconducting high field magnets is one of the pillars of any successful magnet design. Heat losses are inherent to superconductive magnet windings; they have a considerable inpact on the magnet design and its ultimate performance. With respect to their **duration** one can subdivide the heat losses into **steady state** and short, **transient** losses. Steady state losses include the heat input to the coolant by thermal conduction and radiation between parts of the magnet at ambient (or intermediate) temperature and those at liquid helium temperature, heat deposited in the windings due to beam losses, ohmic losses generated in the conductor or cable joints, ac losses due to ripples in the excitation current, and the magnetization and eddy current losses which also appear in dc operated magnets during excitation and de-excitation. These losses can be calculated with reasonable accuracy and cooling systems designed for their safe elimination.

The second group includes short term disturbances resulting in transient heat losses which may trigger a premature quench. Not so long ago these quenches had been considered to be inherent to superconductive high field magnets, which either reached nominal performance after considerable training or failed at lower excitation levels. Based on extensive calculations and experimental results, it is today generally admitted that most of the transient perturbations are of **mechanical origin**. Micromovements caused by the repositioning of strands and cables in the winding are accompanied by sudden flux-jumps and energy release. One of the main tasks for designers of high performance superconducting magnets will consist in developing adequate mechanical structures which will keep the excitation coils under compression, preventing micromovements as much as possible. To this end one can either use epoxy-resin-impregnated coils or use so-called helium-transparent windings, allowing the coolant to penetrate into the coils and absorb instantaneous heat pulses. In both cases one is confronted with the conflicting requirement for compact coils against more efficient heat evacuation.

Various cooling systems can be designed, all based upon the temperature versus entropy chart and the phase diagram of helium, shown in Fig. 6.1a and 6.1b.

The oldest, still widely used system is **pool boiling He I cooling**. The coolant is saturated He I at atmospheric pressure of 0.1 Mpa or 1 bar and a temperature of 4.25 K. The heat losses will evaporate a certain amount of He I at a rate of 2.6×10^3 kJ m^{-3}, which will have to be recycled or replaced. Pool boiling He I cooling is predominantly

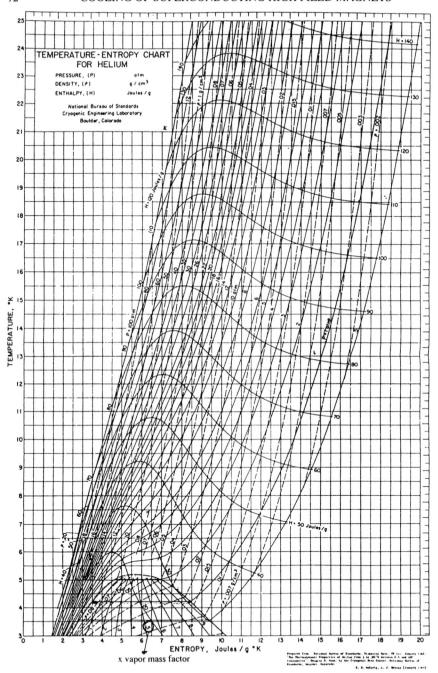

FIG. 6.1a. The T–S diagram of helium.

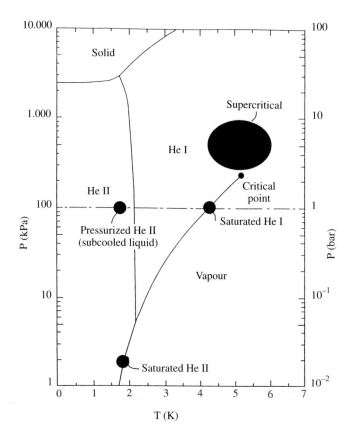

FIG. 6.1b. The phase diagram of helium.

used in test stations and for large individual magnets. The evaporated He I will have to be compensated by refilling the cryostat at regular intervals from the liquefier or dewars. Pool boiling He I is less suited for the cooling of large series of magnets, although it has recently been successfully applied in combination with other cooling modes [1, 2].

Cooling with **single-phase supercritical He I** [3, 4, 5, 6] has been introduced in analogy to forced water flow coolijg of 'classical' magnets. The vapour-free single-phase He I is pushed through cooling ducts in the magnet active parts, or directly through the excitation coils which are then wound with superconductors, embedded in a compact Cu or Al matrix with cooling ducts. In a modified version of 'internally cooled cabled superconductors' (ICCS) the coolant flows through a metallic casing, packed with superconductive strands. Higher heat losses are eliminated by increasing the He mass flow $\dot{m}$ [g s^{-1}], i.e. the velocity v_{He} [ms^{-1}], which in return leads to an increase of the pressure drop through the coils Δp [Mpa] and to fast rising energy losses in the He pumps or refrigerators, providing the required Δp.

The third and most recent cooling mode uses **subcooled superfluid He II** around 1.8–1.9 K, either at a saturation pressure of 20 mbar or preferably at atmospheric pressure. The exceptional heat transfer and heat conduction properties of He II do in certain respects recall the loss-free current conduction in superconductors. Superfluid He cooling will be treated in a separate chapter.

Finally one has to distinguish between the cooling of individual magnets and of large systems of superconductive magnets in high energy physics accelerators and in future fusion machines where dynamic He I or He II cooling is required. In certain cases mixed systems of forced flow single-phase and of pool boiling He I or of saturated and atmospheric pressure operated superfluid He II had been successfully developed. Some of the cooling modes for large magnet systems are described in Chapter 11.

6.1.1   *Parameters for the heat equations*

For a qualitative and numerical analysis of liquid He based cooling systems the heat equations have to be solved, whereby one has to distinguish between heat transport through a medium, conductor or coolant and heat transfer between them. The latter is taken into account by an adequate heat transfer coefficient. To this end the main parameters for the coolant and the conductor have to be introduced. Analytical solutions of reasonable accuracy can be obtained if one assumes that certain parameters do not depend on time, length, or temperature. As a next step elaborate computer programs are introduced for step by step solutions to the heat equations, taking into account also the variation and dependence of each parameter.

The main relevant data concerning He I are summarized in Technical Note 631 of the US National Bureau of Standards [7]. As to the conductor or in general the heat generating part, the extensive compilation of data by H. Brechna [8] on low temperature properties of materials used in superconductive magnets is mentioned. In order to help the reader or designer of superconductiong magnets, some of the data will be reproduced. The main parameters of interest are the conductor matrix resistivity $\rho\,[\Omega\,\mathrm{m}]$, specific heat $C\,[\mathrm{W\,s\,m^{-3}\,K^{-1}}]$, thermal conductivity $K\,[\mathrm{W\,m^{-1}\,K^{-1}}]$, the heat transfer coefficient $h\,[\mathrm{W\,m^{-2}\,K^{-1}}]$, and the heat flux $Q\,[\mathrm{W\,m^{-2}}]$. ρ, C, K depend upon temperature and magnetic induction; Q, h on the cooling mode.

Figures 6.2 and 6.3 show the copper resistivities of different RRR ratios at different magnetic inductions B, while Fig. 6.4 gives the resistivity of high purity aluminium 1100 (0). Figure 6.5 shows the specific heat C of Cu, Al, NbTi, Nb$_3$Sn, and bronze, while Figures 6.6 and 6.7 give the thermal conductivities K for Cu and Al 1100 (0).

6.2   Pool boiling He I cooling

6.2.1   *Steady state and transient heat transfer coefficients*

To investigate the steady state and transient behaviour of pool boiling He I cooling, the heat fluxes Q and Q_t and heat transfer coefficients h, h_t have to be determined. The steady state Q values had been measured by various authors, as shown in Fig. 6.8. The

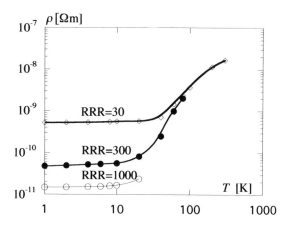

FIG. 6.2. Dependence of copper resistivity upon temperature.

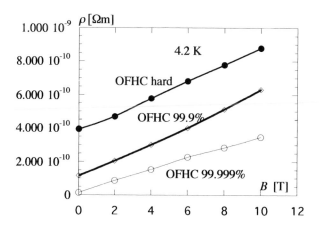

FIG. 6.3. Magnetoresistance of copper for different degrees of annealing.

sizeable spread of results can be explained by differences in the test setup; heat transfer
to pool boiling He I depends on the conductor material and surface conditions and on
the size, inclination and geometry of the cooling channels relative to the conductor.

Figure 6.10 shows a typical heat transfer pattern with the three regions representing
cooling by convection, nucleate boiling, and film boiling. The first region has a low h
defined by the dimensionless Nusselt number

$$Nu = \frac{h d_h}{K} = \frac{Q d_h}{\Delta T K} \tag{6.1}$$

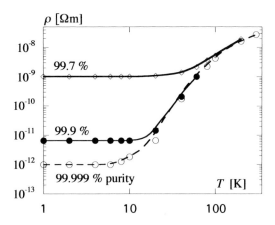

FIG. 6.4. Resistivity of high purity aluminium.

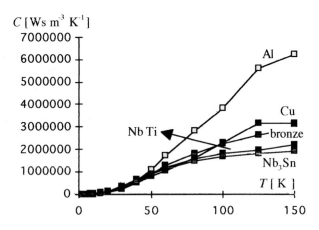

FIG. 6.5. Specific heat dependence upon temperature $C = f(T)$ of some metals, alloys and superconductors.

with d_h the hydraulic diameter of the cooling duct. At higher heat fluxes bubbles appear in the heated surface cavities, forming a $\approx 10\,\mu$m vapour layer. The bubbles either stay within this layer or grow beyond it. This is the **nucleate boiling** region. The third **film boiling** occurs at large heat fluxes above $Q \geq 1000\,\mathrm{W\,m^{-2}}$. Beyond a critical heat flux Q_{cr} defining maximum film boiling, a large temperature jump of $\Delta T \approx 10$ K is observed due to the formation of a layer of insulating He gas on the conductor surface. Decreasing the heat flux, ΔT is reduced to values in the nucleate boiling region. Figure 6.8 shows typical $Q = f(\Delta T)$ curves, measured on flat copper surfaces; despite

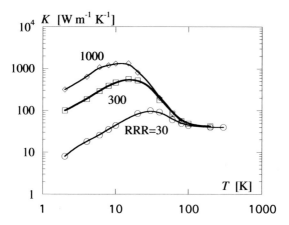

FIG. 6.6. Thermal conductivity of copper.

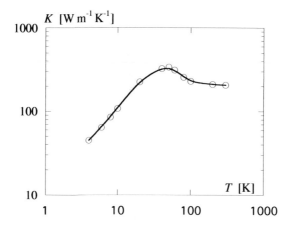

FIG. 6.7. Thermal conductivity of aluminium.

the spread in the results, one can assume for nucleate boiling an average heat transfer coefficient of $600\,\mathrm{W\,m^{-2}\,K^{-1}} \le 1000\,\mathrm{W\,m^{-2}\,K^{-1}}$. Schmidt [9] gives a conservative approximation for $Q\,[\mathrm{W\,m^{-2}}] = f(\Delta T)$:

$$Q = 10^4 \times \Delta T^{2.5}. \tag{6.2}$$

Heat transfer coefficients for the film boiling region are considerably lower and values of

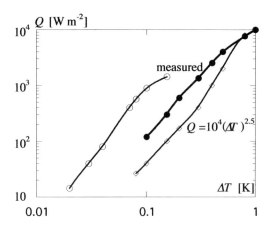

FIG. 6.8. Steady state heat fluxes in pool boiling helium measured for flat copper conductors.

$20\,\mathrm{mW\,m^{-2}\,K^{-1}} \leq h_f \leq 30\mathrm{mW\,m^{-2}\,K^{-1}}$ can be assumed for numerical calculations. In numerical analysis such as the heat equation the variable heat transfer coefficients shown in Figs 6.8 and 6.9 can be approximated by a constant h, as shown in Fig. 6.10.

6.2.2   Heat transfer coefficients for forced flow, two-phase He I

In specific cases forced flow two-phase He I is used for the cooling of superconducting magnets [10]. Figure 6.11 shows the heat flux Q as a function of ΔT for different mass velocities v_g [kg m^{-2} s^{-1}]; beyond a certain ΔT a sharp flux increase is observed. The dependence of this burnout flux Q_b on v_g is shown in Fig. 6.12. An attempt has been made to correlate the heat transfer with the Nusselt number:

$$Nu = 0.024(\mathrm{Re})^{0.9}(\mathrm{Pr})^{0.4}\left(\frac{T_0}{T}\right)^{0.716}, \qquad (6.3)$$

Re and Pr being the dimensionless Reynolds and Prandtl numbers. The experiments do, in fact, agree with a corrected Nusselt number Nu_c by the Martinelli parameter x_{tt}, which takes the He I quality factor x into account (x being the percentage evaporated mass of helium):

$$x_{tt} = \left(\frac{1-x}{x}\right)^{0.9}\left(\frac{\rho_v}{\rho_l}\right)^{0.5}\left(\frac{\nu_l}{\nu_v}\right)^{0.1} \qquad (6.4)$$

ρ_l, ρ_v, ν_l, ν_v being the densities and viscosities of the liquid and vapour parts. The experimentally determined Nu_c is then:

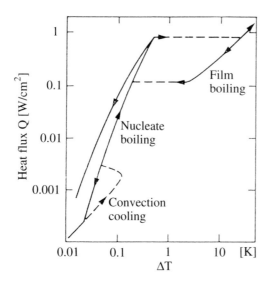

FIG. 6.9. Typical heat transfer characteristic for pool boiling helium.

$$Nu_c = Nu \, x_{tt}^{-0.385} \tag{6.5}$$

6.2.3 *Transient heat transfer coefficients*

The development of fast low temperature measuring devices [9] allowed the transient heat transfer mechanism down to the μs region to be explored. Figure 6.13 shows the $\Delta T = f(t)$ curves, measured on copper samples exposed to a sudden heat pulse for different heat fluxes Q. Heat fluxes well above the steady state nucleate boiling maxima have been measured. The fast temperature rise after the onset time t_f is explained by the beginning of film boiling. For $t \ll t_f$ the curves in Fig. 6.14 can be approximated by the relation:

$$\Delta T = \Delta T_0 + \frac{Q}{h_k} \tag{6.6}$$

with $0.13 \text{ K} < \Delta T_0 < 0.25 \text{ K}$ and a transient heat transfer coefficient of $h_k = 10^5 - 1.5 \times 10^5 \text{ W m}^{-2} \text{ K}^{-1}$. In the film boiling region ten times smaller values had been measured. The parameter h_k^{-1} is identified as the **Kapitza resistance**, characterized by a T^3 temperature dependence below the λ point. The measured values are $1300 \text{ W m}^{-2} \text{ K}^{-4} < \frac{h_k}{T^3} < 2000 \text{ W m}^{-2} \text{ K}^{-4}$. Another important parameter concerning heat transfer in pool boiling He I is the onset time t_f, related to the maximum heat flux Q_m. This can be computed from a simple model, based on the low thermal diffusivity of He I, amounting to $\frac{K}{C} = 4.8 \, 10^{-8} \text{ m}^2 \text{ s}^{-1}$ and the limited thickness of the liquid beneath the

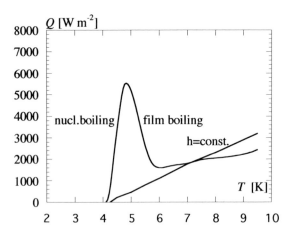

FIG. 6.10. Assumption of constant heat transfer coefficient h for pool boiling helium.

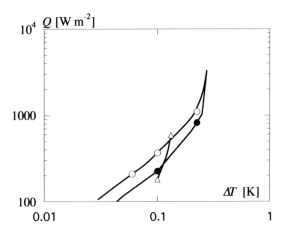

FIG. 6.11. Measured heat fluxes for two-phase He I cooling.

heated surface. Applying a heat pulse to the conductor surface, heat will after a time t have diffused into a layer of thickness Δ [m]:

$$\Delta = \frac{\pi}{2} \left(\frac{K}{C} t \right)^{0.5}. \tag{6.7}$$

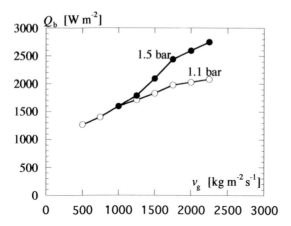

FIG. 6.12. Burnout fluxes in two-phase He I cooling depending upon mass, velocity and pressure.

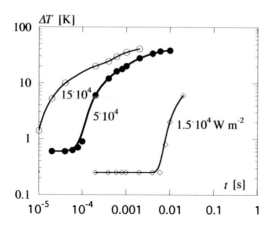

FIG. 6.13. Results of transient heat transfer experiments in pool boiling He I.

Assuming that the transition to film boiling occurs after t_f [s] when sufficient heat $E[\mathrm{W\,s\,m^{-2}}]$ had diffused into the layer Δ to evaporate all the liquid:

$$E = Qt_f = \Delta L \qquad (6.8)$$

$$Q = \frac{\pi}{2} L \left(\frac{K}{Ct_f} \right)^{0.5}. \qquad (6.9)$$

Introducing the value for $L[\mathrm{kJ\,m^{-3}}]$ the heat of vaporization for He I, one obtains:

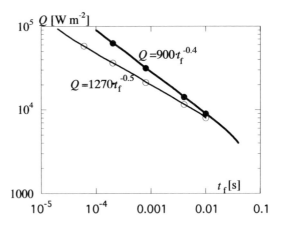

FIG. 6.14. Correlation between 'take-off' time and heat load for He I cooling.

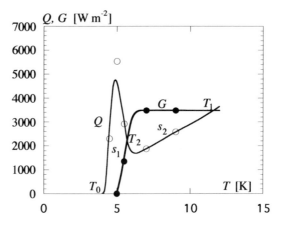

FIG. 6.15. Maddock's stability limit for He I cooling, given by the equal area theorem.

$$Q = 900 \frac{1}{\sqrt{t_f}}. \qquad (6.10)$$

In practical cases the heat flux Q from a surface will be given and the relation $\Delta T(t)$ will have to be determined. According to [9] the critical film boiling heat flux Q_f can be defined:

$$Q_f = 0.032 E^{-0.66} \qquad (6.11)$$

below which $Q < Q_f$; $Q = A(T^4 - T_0^4)$ and above Q_f, $Q = h_f(T - T_0)$, with $A = 300$ W m^{-2} K^{-4} and $h_f = 250$ W m^{-2} K^{-1}. The increased transient Q and h values according to Fig. 6.13, compared to the steady state figures shown in Fig. 6.8, improve the performance of high field and high j_c superconductive magnets.

6.2.4   Heat equations and stability in pool boiling He I

Analytical solutions to the **linear** heat equation for pool boiling **He I** cooling can be obtained assuming that the conductor or cable extends in the x-direction only. The other parameters are the coolant temperature $T_{0.2}$, the conductor, current sharing and critical temperatures, T, T_{cs}, $T_{c.2}$, the thermal conductivity of the stabilizing material K, the average specific heat per unit volume of the composite C, the stabilizer and conductor cross-sections A_{Cu}, A, the cooled perimeter of the conductor P[m], the fraction λ of the superconductor in the composite, the transport and critical currents I_t, I_c, the related overall current densities j, j_c, the heat flux Q, and the heat transfer coefficient h. The resistivity of the stabilizing matrix ρ is assumed to be $\rho = 0$ for $T < T_{cs}$ and increasing linearily to ρ_0 between $T_{cs} < T < T_c$. The heat generated per unit surface $G(T)$[W m^{-2}] is then:

$$G(T) = \frac{\rho_0 I^2}{A_{cu} P} \frac{T - T_{cs}}{T_c - T_{cs}} \tag{6.12}$$

and the cooling per unit surface:

$$Q(T) = h(T - T_0) \tag{6.13}$$

The linear heat equation is then:

$$AC(T)\frac{\partial T}{\partial t} = \frac{\partial}{\partial x}\left[A_{Cu}K(T)\frac{\partial T}{\partial x}\right] - P[Q(T) - G(T)]. \tag{6.14}$$

This states that the temperature increase of the conductor at a distance is equal to the heat contribution by thermal conduction and to the difference between the generated ohmic heating and cooling. Analytical solutions can be obtained by assuming constant values for $C(T)$ and $K(T)$ with the usual choice $C = C(T_c)$, $K = K(T_0)$.

We shall now look for solutions which yield different stability criteria. In the first **steady state** solution long normal conducting zones with $K(T) = 0$ and $AC(T)\frac{\partial T}{\partial t} = 0$ are assumed. Stability is obtained when cooling is superior to the heat generated in the normal conducting zones, $Q(T) \geq G(T)$:

$$\alpha = \frac{G(T)}{Q(T)} = \frac{\rho_0 I^2}{A_{Cu} P h (T_c - T_0)} \leq 1. \tag{6.15}$$

α is the Stekly parameter [11] indicating that the conductor is still efficiently cooled when the full nominal current I_n flows through the stabilizer. For this conservative case one can choose 600W m^{-2} K^{-1} $< h < 1000$ W m^{-2} K^{-1}.

Maddock [12] has improved Stekly's concept by also considering heat conduction $K(T)$ from the normal zones, but neglecting any time variation. According to Fig. 6.15 the two conditions for a stationary solution are $Q(T) \geq G(T)$; $\frac{\partial Q}{\partial T} \geq \frac{\partial G}{\partial T}$. This is the **equal area theorem** which refers to the areas enclosed by the $G(T)$ and $Q(T)$ curves. The stable points are T_0 and T_2, but not T_1. In the transition region between T_0 and T_1 heat conduction provides the balance in accordance with the equal area theorem:

$$\int_{T_0}^{T_1} K(T)[Q(T) - G(T)]\,dT = 0. \tag{6.16}$$

In that case the allowed heat flux is increased to $Q_{ea} \approx 3000 \text{ W m}^{-2}$.

The equal area theorem leads to the concept of **minimum propagating zones (MPZ)**, which states that under specific steady state conditions short lengths of uncooled zones can exist with temperatures of $T > T_c$. The simplest physical explanation for an MPZ is the equilibrium under certain conditions between the excess of heat, the difference between the ohmic losses and cooling in that normal conducting zone, and the axial or longitudinal heat conduction towards the superconductive part of the conductor. In the centre of an MPZ the net heat flux is zero because of symmetry conditions and $\left(\frac{\partial T}{\partial x} = 0\right)_{x=0}$. Meuris [13] has analysed the stability criteria for an uncooled conductor of length l. Starting from the time independent heat equation

$$KA\frac{d^2 T}{dx^2} - P[Q(T) - G(T)] = 0; \ x \geq l \tag{6.17}$$

and $Q(T) = 0$ for $x \leq l$, $T_{x \to \infty} = T_0$ and $\left(\frac{dT}{dx}\right)_{x=0}$, one obtains zero, one or two solutions, depending on h, $G(T)$, l. The solutions are shown in Fig. 6.16a with the following normalized parameters:

$$x_s = \left(\frac{hP}{KA}\right)^{0.5} l \tag{6.18}$$

$$\alpha_i = \frac{\rho j^2 A}{hP(T_c - T_0)} \tag{6.19}$$

$$\theta = \frac{T - T_0}{T - T_c}. \tag{6.20}$$

The area marked 'no solution' has no non-trivial solution and whatever the disturbance, the stability in this region is global, and the conductor will always return to the super-conductive state. The limiting curve determines the cold end recovery current I_{r0}; for an uncooled region of zero length one obtains:

$$I_{r0} = \sqrt{\frac{2hP(T_c - T_0)}{\dfrac{\rho}{A}}} \tag{6.21}$$

which corresponds to the equal area theorem of Maddock. When l increases, I_r decreases according to

$$I_r = \sqrt{\frac{2hP(T_c - T_0)}{l^2hP} \frac{A}{\rho}} \quad \text{for } l < \sqrt{\frac{KA}{Ph}} \qquad (6.22)$$

$$I_r = \sqrt{\frac{4hP(T_c - T_0)A}{\rho} \frac{1}{l\left(\frac{hP}{KA}\right)^{0.5} + 1}} \quad \text{for } l > \sqrt{\frac{KA}{Ph}} \qquad (6.23)$$

However, for any uncooled region of length l the maximum recovery current I_{rm} will even at infinite h be limited to:

$$I_{rm} \le \frac{A}{l}\sqrt{\frac{4K(T_c - T_0)}{\rho}}. \qquad (6.24)$$

Conversely, one can define a maximum uncooled region length l_m still meeting the cold end recovery condition

$$l_m = \frac{2}{I_0}\sqrt{\frac{KA^2(T_c - T_0)}{\rho}}. \qquad (6.25)$$

Finally stable normal zones of finite length may exhibit maximum temperatures T_m higher than those resulting from the equal area criterion. For a temperature T_s at the edge of an uncooled region, T_m will amount to

$$T_m = T_s + \frac{\rho j^2 l^2}{2K}. \qquad (6.26)$$

The existence of uncooled stable normal zones had first been observed in superconducting solenoids with vertical cooling channels materialized by vertical insulating spacers between adjacent layers. The described phenomena may also help in cases where for whatever reason the conductor cooling is locally obstructed over short lengths satisfying eqns (6.24) and (6.25).

6.2.5 *Stability conditions in pool boiling He I cooling with time dependent perturbations*

In the previous chapter stability conditions for time independent perturbations were examined. In this chapter stability conditions for *time dependent* energy perturbations will be analysed. To this end the heat equation (6.14) is written in the parametric form, with $\xi = \frac{x}{l}$:

$$\frac{\partial \theta}{\partial \tau} = \frac{\partial^2 \theta}{\partial \xi^2} + (bw - H)\theta = \frac{\partial^2 \theta}{\partial \xi^2} + \beta\theta = 0 \qquad (6.27)$$

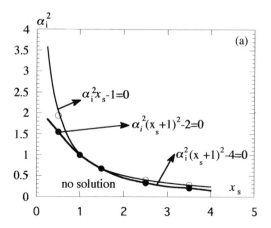

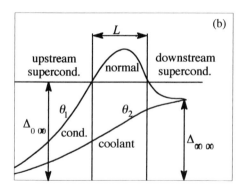

FIG. 6.16. (a) Stability conditions for uncooled regions of a conductor in pool boiling He I; (b) quasi-steady state temperature profiles in forced flow, supercritical He I cooled superconductors.

with

$$\beta = \frac{\rho(\theta)}{\rho_0} w - H = bw - H,$$ (6.28)

normalized time

$$\tau = \frac{K A_{Cu} t}{C A l^2}$$ (6.29)

the normalized heat generation

$$w = \frac{\rho I^2 l^2}{A_{Cu}^2 K (T_c - T_0)} \tag{6.30}$$

and normalized cooling

$$H = \frac{hPl^2}{KA_{Cu}}. \tag{6.31}$$

The normalized length and temperature are given by eqns (6.18) and (6.20). Bejan and Tien [14] have worked out a solution for a δ-function like heat input E_0[W s], introduced in the normalized form

$$\epsilon = \frac{E_0}{2AC(T_c - T_0)\sqrt{\pi\kappa}} \tag{6.32}$$

with κ [m^2 s^{-1}]

$$\kappa = \frac{A_{Cu}K}{AC}. \tag{6.33}$$

The solution to eqn (6.27) is

$$\theta(x, t) = \exp(\beta\tau)\frac{\epsilon}{\sqrt{t}} \exp\left(-\frac{x^2}{4\kappa t}\right). \tag{6.34}$$

The first exponential term corresponds to the balance between heat generation bw and cooling H, while the second term stands for the spatial repartition of the initial heat pulse E_0. For a negative exponent in the first term, there will always be recovery; for a positive exponent $\beta\tau$, recovery is possible in the case of a small enough second term. The extension in time of the normal zone to left and right is given by

$$l_n = 2\left\{4\kappa t\left[\ln\frac{\epsilon}{\sqrt{t}} + \left(\frac{\rho}{C(T_c - T_0)}\left(\frac{I}{A}\right)^2 - \frac{hP}{CA}\right)t - \ln(\theta_c)\right]\right\}^{0.5} \tag{6.35}$$

The recovery of the normal zone is conditioned by a finite temperature maximum of the heat pulse in time [15, 16]

$$\ln\theta_c = \ln\left(\frac{\epsilon}{\sqrt{t}}\right) + 2\left[\frac{\rho}{C(T_c - T_0)}\left(\frac{I}{A}\right)^2 - \frac{hP}{CA}\right]t - 0.5. \tag{6.36}$$

Pasztor and Schmidt [17] have neglected heat generation and cooling during the δ-function like energy perturbation E_0; eqn (6.34) then simplifies to

$$\theta(x, t) = \frac{\epsilon}{\sqrt{t}} \exp\left(-\frac{x^2}{4\kappa t}\right). \tag{6.37}$$

The conductor length l_n where $T > T_c$ is then

$$l_n = 4\left[\kappa t \ln\left(\frac{\epsilon}{\sqrt{t}}\right)\right]^{0.5}. \tag{6.38}$$

The normal zone grows to a maximum length l_m and disappears after the recovery time t_r. If the additional heat ΔE due to ohmic losses in the normal zone is small compared to E_0 such as $\Delta E \leq 0.5 E_0$ one obtains

$$E_0 \leq 4.4 AC(T_c - T_0)l_0 \tag{6.39}$$

with the characteristic length l_0. The maximum normal zone length and recovery time are then

$$l_{nm} = 2.1 l_0 \tag{6.40}$$

$$t_{rm} = \frac{1.5 l_0^2}{\kappa}. \tag{6.41}$$

Extended short duration perturbations have been treated by Nick [18], who applied a transient heat transfer coefficient h_{tr}. The duration of the perturbation depends on the heat flux $Q_{tr}[\text{W m}^{-2}]$ or on the total energy E, transferred per surface S. The condition for zero point stability is then:

$$\frac{Q_{tr} E}{S} \leq \delta^2 \tag{6.42}$$

with $\delta \approx 100[\text{W m}^{-2}\text{s}^{0.5}]$. Due to the high h_{tr} heat conduction can be neglected and a transient Stekly parameter α_{tr} defined:

$$\alpha_{tr} = \frac{\rho_0 I^2}{A_{Cu} P h_{tr}(T_c - T_0)}. \tag{6.43}$$

Recovery is obtained for $\alpha_{tr} < 1$. The recovery time is

$$t_r = \frac{AC}{P h_{tr}(1 - \alpha_{tr})} \tag{6.44}$$

and the recovery energy density limit $e_0[\text{Ws m}^{-3}]$ is

$$e_0 = \delta\left[\frac{PC(1 - \alpha_{tr})}{A h_{tr}}\right]^{0.5}. \tag{6.45}$$

6.3 Forced flow cooling with single-phase supercritical helium

The advantages of this cooling mode were first stressed by Holm [3]. At reasonable flow rates of the coolant, equal or slightly higher heat transfer coefficients than for pool boiling He I can be obtained. In the film boiling onset region the decrease of h is less pronounced, and flow instabilities due to vapour density differences are smaller. The design of large individual superconducting magnets can be simplified by using a vacuum tank instead of a helium container in a vacuum vessel. The coils can be wound with compact hollow conductors, adequately insulated and impregnated with epoxy resin, which will considerably improve the dielectric strength.

The disadvantages are linked to the required pressure drop Δp and to the related power consumption in the helium pumps or refrigerators. The heat transfer coefficient h is given by eqns (6.1) and (6.3) [5, 6]:

$$h = 0.024 \frac{K}{d_h} (\text{Pr})^{0.4} (\text{Re})^{0.8} \left(\frac{T_{He}}{T_{cond}} \right)^{0.716} \tag{6.46}$$

$$(\text{Re}) = \frac{\dot{m} d_h}{A_{He} v} \tag{6.47}$$

$$(\text{Pr}) = \frac{v C}{K} \tag{6.48}$$

with d_h the hydraulic diameter, $\dot{m}$ [g s^{-1}] and v[g s^{-1} m^{-1}] the helium mass flow and viscosity.

The stability of forced flow cooling with supercritical helium is governed by two partial differential equations, one for the conductor and the second for the coolant. For helium the varying temperature T_{He}, specific heat C_{He} and velocity v_{He} have to be considered. The coupled heat equations are:

$$A C(T) \frac{\partial T}{\partial t} = \frac{\partial}{\partial x} \left[A K(T) \frac{\partial T}{\partial x} \right] - P[Q(T) - G(T)] \tag{6.49}$$

$$A_{He} C_{He} \left(\frac{\partial T_{He}}{\partial t} - v_{He} \frac{\partial T_{He}}{\partial x} \right) = P h(T - T_{He}). \tag{6.50}$$

Greene and Saibel [19] have worked out analytical solutions to eqns (6.49) and (6.50) by introducing a quench origin propagating velocity v_1 and $v_2 = v_{He}$, $k = K A$, $c = C A$, $c_1 = c_{cond}$; $c_2 = c_{He}$, $T_1 = T_{cond}$; $T_2 = T_{He}$. Assuming a quasi-stationary state and neglecting the time dependent variations the above equations can be written as:

$$k \frac{\partial^2 T}{\partial x^2} + c_1 v_1 \frac{\partial T}{\partial x} = P[Q(T_1 - T_2) - G(T)] \tag{6.51}$$

$$c_2 (v_2 - v_1) \frac{\partial T_2}{\partial x} - h P (T_1 - T_2) = 0. \tag{6.52}$$

The solution is given by the following relations:

$$\frac{1 + [1 - \exp(r_1 l_c)]\dfrac{1}{r_1 l_c}}{[1 - \exp(-r_2 l_c)]\dfrac{1}{r_2 l_c} + [1 - \exp(r_1 l_c)]\dfrac{1}{r_1 l_c}} = \left[\dfrac{c_2(v_2 - v_1)}{hP}r_2 + 1\right]\dfrac{r_1}{r_1 - r_2} \tag{6.53a}$$

where $r_1, r_2 [\mathrm{m}^{-1}]$ are given by

$$r_{1,2} = -\frac{1}{2}\left[\frac{hP}{c_2(v_2 - v_1)} + \frac{c_1 v_1}{k}\right]$$
$$\pm \sqrt{\frac{1}{2}\left[\frac{hP}{c_2(v_2 - v_1)} + \frac{c_1 v_1}{k}\right]^2 + \frac{hP}{k}\left[1 - \frac{c_1 v_1}{c_2(v_2 - v_1)}\right]^2}. \tag{6.53b}$$

The quasi-steady state temperature profile is shown in Fig. 6.16b. In the case of a quench three zones can be distinguished, a downstream and upstream superconductive and a central normal zone of critical length l_c. Equation (6.52a) gives the $l_c(v_1)$ dependence for the normal propagating zone. The corresponding relation between conductor heating and cooling for a given v_1 and l_c is:

$$\frac{G(T)}{h(T_c - T_0)} = \frac{r_2}{1 - \exp(-r_2 l_c)}\frac{\left[\dfrac{c_2(v_2 - v_1)}{hP}\right]\left[\dfrac{1 - c_1 v_1}{c_2(v_2 - v_1)}\right]}{\dfrac{r_2}{r_1 - r_2}\left[\dfrac{r_2 c_2(v_2 - v_1)}{hP} + 1\right]}. \tag{6.53c}$$

Equations (6.52b) and (6.52c) determine the stability conditions in internally cooled superconductors with supercritical helium depending on v_2 and T_2, the current and stabilizer. The following conclusions can be drawn: an internally cooled superconductor will be stable for $l < l_c$. The normal zone will then propagate upstreams, contract, and disappear. For $l > l_c$ the normal zone will expand and propagate downstream and stability will then depend upon the conductor stabilization. At low stabilization the upstream edge of the normal zone will propagate downstream, opposite to v_2 and the normal zone can only be cleared by reducing the excitation current. l_c can be found for any configuration, but not when the maximum conductor length l_{nm} becomes accidentally normal. In practical cases it will seldom be possible to design stable cooling systems with $l < l_c$. In the case of a premature quench the helium mass flow $\dot{m}_{He}$ and consequently the heat transfer coefficient can be increased [eqns (6.46) and (6.47)]. Figure 6.17 shows a comparison of various cooling modes, including two cases of forced flow cooling for two Reynold numbers $Re = 10^5$ and 10^6. Increasing Re corresponds to an increase of the mass flow $\dot{m}$ and of the pressure drop due to friction:

$$\frac{\mathrm{d}p}{\mathrm{d}x} = f(Re)\frac{\dot{m}^2}{2\rho_{He} A_{He}^2 d_h} \tag{6.54}$$

with

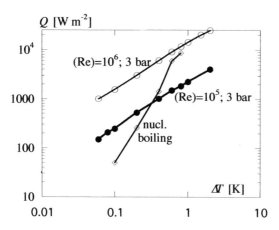

FIG. 6.17. Comparison of heat fluxes in pool boiling He I and in forced flow He I cooled superconductors.

$$f(\text{Re}) = 0.184(\text{Re})^{-0.2}. \qquad (6.55)$$

The increase in heat transfer is thus proportional to $\dot{m}^{0.8}$, and the associated pressure drop to $\dot{m}^{1.8}$. The required pumping power per unit length increases as:

$$\frac{w_p}{l} = \frac{\dot{m}}{\rho_{\text{He}}}\frac{dp}{dx} = f(\text{Re})\frac{\dot{m}^3}{2\rho_{\text{He}}^2 A_{\text{He}}^2 d_h}. \qquad (6.56)$$

It is thus proportional to $\dot{m}^{2.8}$ and to P^{-3}.

The above relations lead to the development of **internally cooled cabled superconductors** (ICCS) [20], where bundles of NbTi or Nb$_3$Sn wires are packed into an outer, usually stainless steel conduit, and cooled by forced flow He I or He II. The sum of the cooling perimeters of the individual wires is then much greater compared to a compact superconductor with the same number of wires or strands. The long, restrained flow paths will give rise to high pressures in the case of a quench, as shown by Dresner, Krafft, Zahn and Ries [5, 6, 21, 22]. A simple expression has been developed for the maximum pressure increase Δp_m [MPa] for unit volume heating q [W m^{-3}] and length of the flow path $2l$

$$\Delta p_m = 0.1\left(\frac{q^2 l^3}{d_h}\right)^{0.36}. \qquad (6.57)$$

Calculated and measured values for Δp_m are shown in Fig. 6.18.

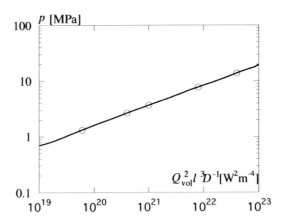

FIG. 6.18. Maximum pressure in forced flow He I-cooled superconductors.

6.4 Superconducting high field magnets cooled with superfluid He II

6.4.1 *General remarks*

During the last two decades the applicaton of superfluid helium or He II as a coolant for superconducting high field magnets has been rapidly expanding. The understanding and collection of experimental data on superfluid helium has been widened; its exceptional thermodynamic properties like high thermal conductivity, high specific heat, and velocity of the loss-free component make this fluid particularily attractive. Following the proposal of Claudet *et al.* [23, 24] to use **pressurized** He II at 0.1 MPa or 1 bar instead of saturated He II at 20–50 mbar one could avoid the inconvenience of using cryogenic systems, operating at subatmospheric pressures. The growing demand for magnets with increasing inductions B and current densities j_c in NbTi and Nb_3Sn also pointed towards operating temperatures below 4.2 K, preferably below 2 K (see Sections 5.2 and 5.5.5). The feasibility and reliability of a large cooling system with pressurized He II was demonstrated by the Tore-II project of an experimental fusion reactor in France [25]. Among the disadvantages with respect to other cooling systems the greater complexity, higher costs, the low dielectric strength of He II and an increased sensitivity to heat input by radiation, conduction, and ohmic losses are mentioned.

6.4.2 *Thermodynamic properties of superfluid He II*

According to the phase diagram shown in Fig. 6.16 superfluid He II is liquid up to pressures of 2.5 MPa or 25 bars. The two liquid phases of He II and He I are separated by the so-called λ-line extending from the low pressure end at 2–5 kPa and 2.172 K to the upper end at 3.023 MPa and 1.763 K; at atmospheric pressure the λ phase transition occurs at 2.158 K. The large specific heat C is shown in Fig. 6.19 with a peak just below the λ point. The particular hydrodynamic behaviour of He II can be described by a two-

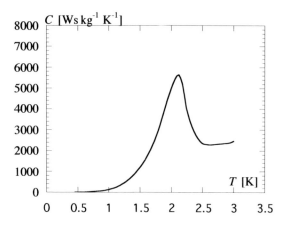

FIG. 6.19. Specific heat of superfluid helium.

fluid model of normal density ρ_n [kg m^{-3}] and superfluid density ρ_s, first introduced by Khalatnikov [26].

The normal component transports the thermal energy and exhibits entropy and viscosity [27]; the superfluid component behaves like an inviscid liquid without entropy carrying no thermal energy. Figure 6.20 shows the temperature dependence of ρ_n and ρ_s. The normal component heat flow is

$$Q = \rho_n S T v_n \qquad (6.58)$$

with S [W s kg^{-1} K^{-1}] the unit mass entropy of the liquid. A temperature gradient $\frac{\delta T}{\delta x}$ in a cooling channel filled with He II will result in forces acting in opposite directions on the normal and superfluid components, making the liquid a very efficient heat conductor. When heat is transported from a heat source, a conductor, to and through superfluid He II, dissipative mutual friction losses occur in the normal component in the direction from the source to the liquid, while the superfluid component flows with great speed up to $v_s = 20$ m s^{-1} towards the heat source.

Heat transport in an assumed unidimensional channel of length $x = l$, filled with He II and transporting a heat flux Q [W m^{-2}] between a heat source at temperature T_w and T_{HeII} is given by the **Gorter–Mellink** equation:

$$-\frac{dT}{dx} = f(T)Q^{3.4}. \qquad (6.59)$$

Some authors [28, 29] use the same equation but with a Q^3 heat flux dependence. $f^{-1}(T)$ is the inverse heat conductivity function, shown in Fig. 6.21 given by

$$f^{-1}(T) = \frac{\rho_s^3 S^4 T^3}{A_g \rho_n} \qquad (6.60)$$

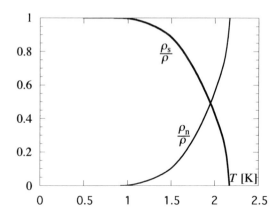

FIG. 6.20. Fractions of normal and superfluid density in superfluid He II.

with a maximum of 1.5×10^{13} [$W^3 \, m^{-5} \, K^{-1}$] at 1.9 K. A_g [$m \, s \, kg^{-1}$] is the Gorter–Mellink mutual friction parameter [27], shown in Fig. 6.22. Integrating eqn (6.59) over the channel length l and between T_w and T_{HeII} yields:

$$f^{-1}(T) \bullet (T_w - T_{HeII}) = X(T_w) - X(T_{He}) = lQ^{3.4} \qquad (6.61)$$

Figure 6.23 shows the graphical solution to eqn (6.60) which allows one to determine one missing parameter. T_w must evidently remain below $T_w < T_\lambda$. Figure 6.24 shows another presentation of eqn (6.60), yielding the critical channel length l_c for a given Q and the parameter T_{He}.

6.4.3 Steady state heat transfer in superfluid helium

In analogy to the previously described He I cooling modes, heat is now transferred from a superconductor to the surrounding He II. A distinct temperature difference is associated to this process, which is greater than the temperature gradients in bulk He II, described in Section 6.4.2 [27, 28, 30]. The temperature jump due to this heat transfer is related to the process of energy exchange between the elementary excitations in the solid conductor and in He II. Khalatnikov [26] attributes this phenomenon to an acoustic mismatch between the two media, now known as the Kapitza conductance which can be characterized by a heat transfer coefficient h_k [$W \, m^{-2} \, K^{-1}$]. In accordance with Figs 6.25 and 6.26, h_k values of 1000–2000 [$W \, m^{-2} \, K^{-1}$] can be expected at small temperature differences of $\Delta T < 0.3$ K, which are hardly superior to the values obtained for He I cooling. At higher heat fluxes of $Q \geq 10^4$ [$W \, m^{-2}$], h_k exhibits an exponential temperature dependence T^n with $3 < n < 4$. Values of $1000 < h_k < 8000$ have been measured for bare copper surfaces in He II at 1.9 K [29]; the spread of results is due to different treatments of the copper surfaces. The general expression for the heat transfer, governed by the Kapitza conductance, can be written as:

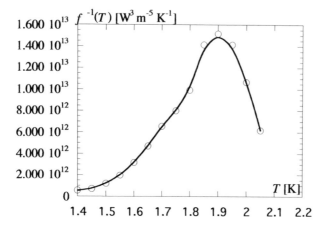

FIG. 6.21. The inverse heat conductivity function f^{-1} of superfluid helium.

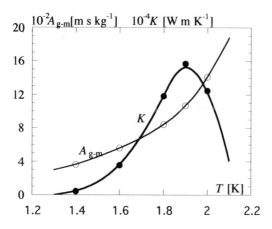

FIG. 6.22. The Gorter–Mellink friction parameter and heat conductivity of superfluid He II.

$$Q_k = a\left(T_w^n - T_{HeII}^n\right); \quad 3 < n < 4 \tag{6.62}$$

For small ΔT of a few 0.1 K one obtains for h_k

$$h_k = 4aT_w^3 \tag{6.63}$$

with $\frac{h_k}{T_w^3} \approx 1\text{–}2$ [W m^{-2} K^{-4}]. The maximum heat flux Q_m is reached when a layer of He I vapour or liquid, or of both is formed. The layer insulates the heat source from

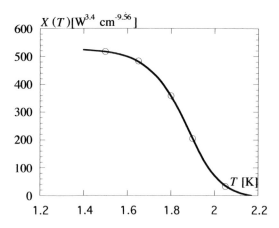

FIG. 6.23. Chart for solving the Gorter–Mellink heat equation $Q^{3.4}L = X(T_w) - X(T_c)$.

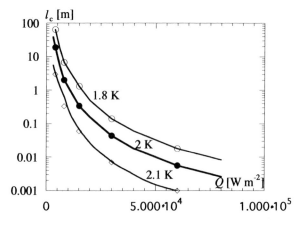

FIG. 6.24. Dependence of critical channel length upon heat flux and He II bath temperature.

the He II and allows the warm conductor surface temperature to increase by an order of magnitude, as shown in Fig. 6.27. The maximum heat flux is then

$$Q_m \approx 490 \times T_w^3. \tag{6.64}$$

Due to the spread in the measured Q and h_k values it is advisable to perform measurements on models representing closely the conditions in He II cooled superconduct-

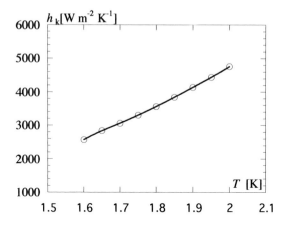

FIG. 6.25. Average steady state heat transfer coefficient, measured in superfluid He II.

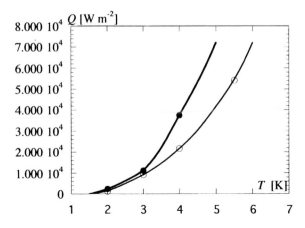

FIG. 6.26. Steady state heat transfer from copper to the Kapitza region of He II at large
fluxes.

ing magnet coils.

Transient heat transfer has been investigated assuming step-function-like heat
pulses and experiments have been carried out by a number of authors [29, 31, 32, 33,
34]. The phenomenon is today reasonably well understood. Here He II has considerable
advantages compared to other cooling modes. Time-dependent heat transfer in He II is
explained by the entropy theorem [29]. The theorem could be formulated after experiments had shown that large heat fluxes of the order of $Q \approx 40\,000$ [W m^{-2}] and pulse

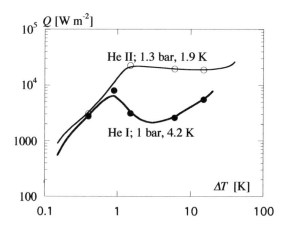

FIG. 6.27. Steady state film boiling heat fluxes in He I and He II, depending upon ΔT.

durations of 0.35 s were absorbed by the He II without onset of film boiling. Due to the high diffusivity of He II, amounting to 0.1–1 $[\text{m}^2\text{s}^{-1}]$, heat penetrates deeply into the bulk of the fluid, while the high specific heat C allows the absorption of large energies at small ΔT. A rigorous analysis of the transient heat transfer and of the stability conditions for He II cooled superconductors has been carried out by Meuris [32]. The conductor configuration with the assumed single-directional flow as shown in Fig. 6.28 has been examined for the two cases of closed and open end channels. The analysis is based on the Gorter–Mellink relation (6.59) and on the energy conservation equation in a stationary fluid:

$$C(T)\frac{\delta T}{\delta t} = -\frac{\delta q}{\delta x}. \tag{6.65}$$

The following cases have been examined: the heat pulse q per unit fluid cross-section is small and the He II temperature is $T < T_\lambda$. The heat transfer is governed by the time-independent Kapitza conductance. In the second case the He II temperature at the conductor is $T \approx T_\lambda$. A thin layer of He I is then formed between the conductor and the fluid. An energy difference appears between the conductor plus He I layer and the He II, which is cancelled when the energy excess in He I is absorbed by the He II. To examine this case eqns (6.59) and (6.65) were solved for the boundary conditions $T = T_\lambda$ for $x = 0$ and $T = T_{\text{He II}}$ for $x = l$ in the case of an open channel, or $q(x = l) = 0$ for a closed channel end. The results are shown in Fig. 6.29; they are presented in terms of the enthalpy change over the channel length between T_λ and T as a function of the Gorter–Mellink parameter $\frac{\Delta E(T_\lambda,T)}{l} = f(ql^{\frac{1}{34}})$. The stability of a superconductor, exposed to low and high power disturbances of **short** duration, can now be determined. The low power disturbance is characterized by the conductor energy per unit volume e_c $[\text{Ws m}^{-3}]$ and the pulse duration t, where in agreement with Fig. 6.30,

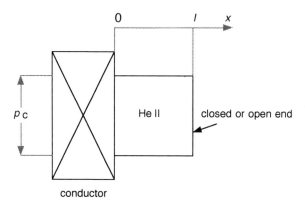

FIG. 6.28. Configuration for investigating the transient heat transfer to superfluid He II.

$$\frac{e_c A_c}{p_c l} < \frac{\Delta E_0}{l} \left(q_0 l^{\frac{1}{3.4}} \right) \tag{6.66}$$

with A_c and p_c the cross-section and wetted perimeter of the conductor, and the energy absorbed by the He II $\frac{\Delta E_0}{l}$ for a step-function-like heat flux

$$q_0 = \frac{e_c A_c}{p_c t}. \tag{6.67}$$

The superconductor will not go normal provided its temperature remains below $T < T_c$:

$$\frac{e_c}{t} = \frac{q_0 p_c}{A_c} = \rho j^2 < a \left(T_c^{3.4} - T_{\text{He II}}^{3.4} \right). \tag{6.68}$$

A short duration, high energy disturbance is characterized by $\frac{e_c A_c}{p_c l} > \frac{\Delta E_0}{l}$ and by $\Delta E_0 \ll \Delta E(T, T_\lambda)$ (short duration). The temperature of the conductor subject to such an energy pulse will rise adiabatically to an initial temperature $T_i > T_c$, while the He II temperature will rapidly increase to T_λ. The instantaneous temperature of the conductor is given by

$$C_c \frac{dT}{dt} = \rho j^2 - \frac{p}{A} q_{\text{He II}}(t) \tag{6.69}$$

with $q_{\text{He II}}$ the heat flux per He II channel cross-section. Integrating eqn (6.69) one obtains for the energy balance per unit conductor volume:

$$e_c + \rho j^2 = e_{\text{He II}} \frac{p}{A} + \int_{T\text{He II}}^{T} C_c dT \tag{6.70}$$

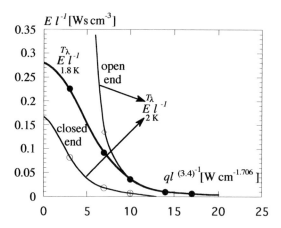

FIG. 6.29. Energy absorbed by an open or closed end He II channel before T_λ is reached at the hot end.

with e_{He} the energy absorbed by the He II and the layer of He I. Conductor recovery and the disappearance of the He I layer will occur after a time t_r, when the total energy will be absorbed by He II:

$$\frac{e_c A_c}{p_c t l} + \rho j^2 \frac{A_c}{l p_c} = \frac{\Delta E(t_r)}{l}. \tag{6.71}$$

Equation (6.71) has been solved graphically for closed and open channel configurations; the results are shown in parametric form in Figs 6.32 and 6.33 with:

$$\Delta E = \frac{\Delta E}{l}; t' = \frac{t}{l^{1+\frac{1}{3.4}}}; x' = \frac{x}{l}. \tag{6.72}$$

The common point to curves G and H representing the total energy in the conductor and the energy absorbed by the He II defines the critical recovery time t'_{rc} beyond which there is no recovery. To determine the critical energy per unit conductor volume $e_c(t_{rc})$, one first determines t'_{rc} by equating the He II flux to the joule heat flux:

$$\frac{A_c}{p_c} \rho j^2 l^{\frac{1}{3.4}} = \dot{q}(t'_{rc}). \tag{6.73}$$

From Fig. 6.31 or 6.32 $\Delta E' = \frac{\Delta E}{l} = f(t_{rc})$ can be determined; the critical conductor energy is then:

$$e_c(t_{rc}) = \left(\frac{\Delta E'}{l} \frac{p_c}{A_c} - \rho j^2\right) t' l^{1+\frac{1}{3.4}}. \tag{6.74}$$

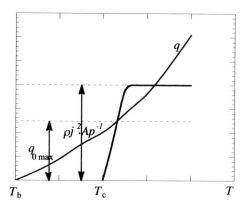

FIG. 6.30. Stability in He II at low power disturbances.

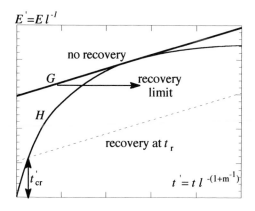

FIG. 6.31. Recovery conditions in a closed end channel at constant current; short perturbation.

Experimental results on the transient stability of He II cooled superconductors show that maximum heat fluxes Q_m and heat energies per unit conductor volume e_{cm} are by at least an order of magnitude higher than for He I cooling [31, 32, 33, 34].

6.4.4 *Heat transport through the insulation of superconducting cables cooled with superfluid He II*

So far steady state and transient heat conduction in He II and heat transfer from a bare conductor surface to the coolant have been examined. In practice, however, superconductive high field magnets will have insulated conductors or cables. In order to benefit

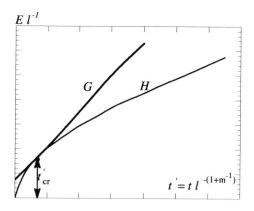

FIG. 6.32. Recovery conditions for open end channel at constant current; short, strong perturbation.

from the considerable advantages of He II cooling one will be well advised to develop configurations of insulated cables which will allow an at least partial penetration of the He II into and through the insulation to the conductor surface. Such geometries are called He II permeable insulations. Since the magnet system of the next large supercollider of CERN, the LHC [35], will be cooled by He II at 1.9 K and 1 bar, it became necessary to study the heat transfer properties of the superconducting cables and insulation for different heat loads. Extensive experiments have been performed by CERN and CEN at Saclay, France [36, 37] with the aim of verifying whether the cable insulation will allow a safe transfer of the steady state heat losses to the coolant. The dominant loss component are the proton beam losses, which in certain magnet coils reach values of 1.0–1.5×10^4 [W m^{-3}]. The maximum temperature increase ΔT should be limited to 0.2 K, in order not to exceed the transition temperature T_λ. Figure 6.33 shows an experimental setup of a stack of insulated cable samples. The samples consist of machined stainless steel bars; the central bar is electrically heated. The two-composite insulation consists typically of two overlapped layers of a 25μm polyimide (or kapton) tape, providing the electrical insulation and of an epoxy resin preimpregnated glass tape with a gap of 2–4 mm for He II penetration. A typical $\Delta T = f(Q_0)$ curve for a heated insulated sample is shown in Fig. 6.34. Up to $T \leq T_\lambda$ only the He II in contact with the conductor will change phase. In the intermediate phase all the He II in the insulation will have attained T_λ. In the asymptotic part no He II will be left in the insulation. Experiments have further shown that this type of heat transport cannot be described by the Gorter–Mellink heat equation or by simple heat conduction, but rather by the relation

$$\Delta T = \alpha Q^n \approx \alpha Q^{1.4}. \tag{6.75}$$

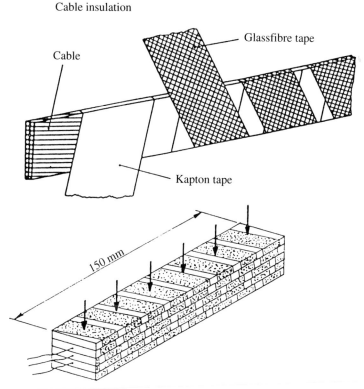

Cable insulation

Glassfibre tape

Cable

Kapton tape

150 mm

FIG. 6.33. Test sample of insulated and heated conductors in superfluid He II. (Courtesy: CERN, Geneva, and DAPHNIA/STCM, CEA/Saclay, France.)

Figure 6.34 shows the two temperature ranges for the central conductor of the setup according to Fig. 6.33. In the first range $\delta T < \Delta T_\lambda$ exhibits a kind of mixed behaviour between conductance and a 'superconductive' regime of the Gorter–Mellink type. The second region is characterized by $\Delta T > \Delta T_\lambda$ and $Q > Q_{cr}(\Delta T_\lambda)$. The critical thermal flux Q_{cr} depends upon the composition and geometry of the polyimide plus fibreglass insulation, as shown in Fig. 6.35. The influence of increasing the fibreglass spacing is shown in Fig. 6.37a. Considerable improvement was obtained by using a preimpregnated glass web plus polyimide, as shown by the two curves on the right side in Fig. 6.36a. In general heat transfer from insulated cables to He II will depend on the longitudinal and transverse cooling channels and on the helium volume in the cable and insulation. An interesting comparison of the heat transport properties, measured on an insulated conductor sample in He II at $T < T_\lambda$ and in He I at $T = 3$ K, both at 1 bar is shown in Fig. 6.36b. At 1.7 K $< T_{He\,II} < 1.9$ K a thermal resistance of ≈ 4 K W^{-1} was measured, compared to 14 W K^{-1} at $T_{He\,I} = 3$ K.

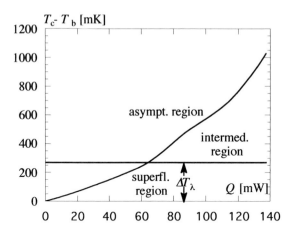

FIG. 6.34. Central conductor temperature increase as a function of the applied heat.

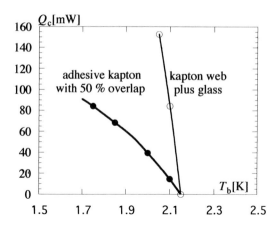

FIG. 6.35. Critical heat fluxes Q_c of samples with different insulations.

6.4.5 *Cryostats for magnet tests in superfluid He II at 1 bar*

An important cryogenic element for cooling tests in He II at atmospheric pressure is the Claudet bath or cryostat [24]. The cooling system of the experimental fusion reactor Tore-II [25] and for the superconducting magnets of the future large collider of CERN, the LHC [35], are both based on the principle of the Claudet bath. He modified and improved the Roubeau bath [38] which provided He II at atmospheric pressure and a temperature of $T_\lambda - \varepsilon$, with $\varepsilon = 10^{-4}$–10^{-5} K, but not in the optimal range of 1.7–1.9 K. The principle of the Claudet bath is shown in Fig. 6.37. The bath or cryostat

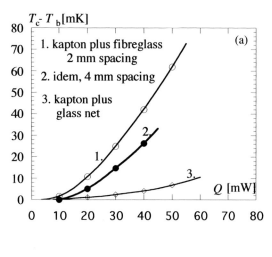

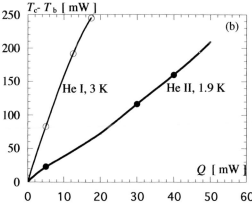

FIG. 6.36. (a) Conductor temperature increase for different insulations; (b) comparison
of heat transfer from an insulated conductor sample to He I and He II.

consists of an upper chamber, filled with He I at 1 bar, whose temperature gradually
decreases from 4.2 K to $T_\lambda - \varepsilon$ at the insulating plate, separating the upper from the lower
chamber. A heat exchanger, placed in the lower chamber is connected to a pumping
system. Initially the entire cryostat is filled with He I at 4.2 K and 1 bar. By pumping
and reducing the pressure in the heat exchanger to and below saturation, the temperature
in the lower chamber will gradually be reduced below T_λ. The heat losses in the lower
or test chamber are evacuated through the pumping line. In addition to the main current
leads, two transfer lines provide the supply of He I and the return of the He I gas. A
Joule–Thompson expansion valve supplies the heat exchanger with He I, tapped from

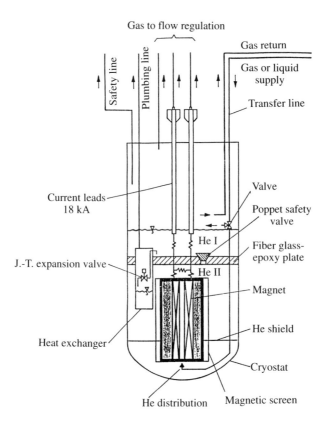

FIG. 6.37. Principle of the He II bath at atmospheric pressure according to Claudet.

the upper chamber. In the Claudet bath the He II temperature can be varied over a large range below T_λ, where superfluid helium exhibits optimal heat transfer and conduction properties.

For readers interested in more details about Helium cooling modes for superconducting magnets the exhaustive book of S.W. Van Sciver [39] is warmly recommended.

7

THE QUENCHING PROCESS IN SUPERCONDUCTING MAGNETS AND THEIR PROTECTION

7.1 General

The transition from the superconductive to the normal state in superconducting magnets and their protection from possible serious consequences of the quenching process have been investigated since the early development of superconducting coils. It has been demonstrated that the electromagnetic and mechanical energies stored in the magnet coils and released during a quench could seriously damage or even destroy them, unless adequate quench protection schemes and devices are implemented.

With the recent development of large-size superconducting magnets wound with high j_c, and high field superconductors of high energy density, the problem of protecting individual or strings of magnets has grown in complexity. With the emergence of large superconductive magnet systems in particle accelerators and colliders one had to extend the notion of quench protection of a single magnet to a string of interconnected magnets. The quenching process and the quench protection analysis comprise a large number of magnet components, which can be divided into **internal** and **external** elements. The first comprise the coil and magnetic iron configurations, the electromagnetic couplings between windings and layers, the superconducting wire, cable and insulation, the metallic parts subject to eddy currents, quench detectors and heaters. External components are the power supplies, switches, discharge resistors, diodes, and other string magnets. In parallel large and sophisticated computer programs for quench analysis had been developed. The early programs like Quench [1] or Tmax [2] had been developed for cases of relatively simple solenoids and required extensive modifications every time one tried to apply them for other, different magnet configurations. Versatile analogue computer programs are nowadays available for any magnet configuration [3, 4, 5, 6]. Only such programs can, in fact, handle the multitude of parameters, many of them time and temperature dependent, and calculate within a reasonable delay a variety of cases, depending upon the assumed quench origin and the variation of certain 'free' parameters. It is stressed that the quench protection system is an integral part of a superconducting magnet (or magnet system), which has to be from the very beginning incorporated into the magnet design. It would be wrong to first design and even build a magnet and then think how to protect it.

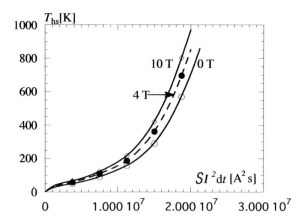

FIG. 7.1. Hot spot temperature dependence upon MIIT values of the TAP twin aperture dipole of CERN.

7.2 Relations determining the quench propagation, 'hot spot' temperature, and the maximum voltages in superconducting magnets

The important parameters to be calculated are the quench propagating velocities v [m s^{-1}], the maximum 'hot spot' temperature T_{hs}, and the maximum voltage U_m [V] between layers or pole windings and with respect to ground. When a superconductive magnet quenches at point z_0 in a winding of assumed linear extension the normal zone will spread upstream and downstream at a velocity v, which can be obtained from the heat equation, already mentioned in Chapter 6. Writing this in a somewhat modified form, one obtains:

$$\frac{\delta}{\delta z}\left(KA\frac{\delta T}{\delta z}\right) - vC_zA\frac{\delta T}{\delta z} - hP(T - T_0) + GA = 0. \tag{7.1}$$

In this equation K and G depend upon T and B, while the heat transfer coefficient h and the specific heat C_z depend on T. Consequently no exact solution exists to eqn (7.1) and computer programs using finite difference methods and stored data on $K(T, B)$, $G(T, B)$, $C(T)$, and $h(T)$ must be used. If one neglects the temperature dependence, one obtains, to a fair approximation, for the normal zone spreading velocity

$$v = \frac{j}{C}\sqrt{\frac{\rho K}{T - T_0}} \cdot \frac{1 - 2y}{\sqrt{1 - y}} \tag{7.2}$$

where

$$y = \frac{hP(T_{sc} - T_0)}{Aj^2\rho}. \tag{7.3}$$

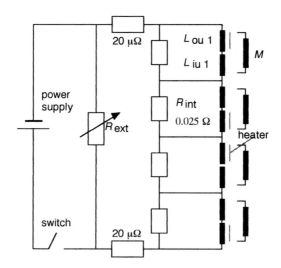

FIG. 7.2. Electrical scheme of a 14 m long, twin aperture LHC dipole of CERN (L_{oul}: outer layer, upper pole; L_{iul}: inner layer, upper pole).

Having calculated the initial velocity v_0 one can calculate its evolution as a function of the excitation or transport current $I_t(t)$:

$$v(t) = v_0 \frac{I_t(t)}{I_0} \left[1 + a \frac{I_t(t)}{I_0} \right] \frac{1}{1+a} \tag{7.4}$$

with I_0 the current at the instant of quenching and a the superconducting wire or cable parameter, where $0.2 < a < 0.5$ [7, 8]. v_0 and v are the longitudinal velocities; to calculate the transverse, turn-to-turn spreading velocity v_{tr} one applies the relation

$$v_{\text{tr}} = v \frac{C_{\text{cable}}}{C_{\text{ins.cable}}} \sqrt{\frac{K_{\text{tr}}}{K_1}}. \tag{7.5}$$

A more detailed expressions for v_{tr} has been worked out by representing the excitation coils by equivalent networks [9]. The hot spot temperature T_{hs} is obtained from the so called MIIT relation

$$\int_0^\infty I^2(t)\mathrm{d}t = A_{\text{Cu}} A_{\text{tot}} \int_0^{T_{\text{max}}} \frac{\Sigma C(T)}{\rho(B,T)} \mathrm{d}T. \tag{7.6}$$

Equation (7.6) compares the accumulation of the incremental ohmic losses, transformed into heat at the quench point, and calculates T_{hs} from the ΣC of all metallic components of a wire or cable. Equation (7.6) is solved in steps, where $I(t) = I[R_n(T), L_n(t),$

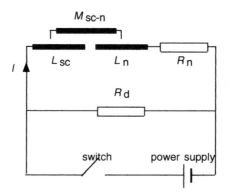

FIG. 7.3. The resulting equivalent electrical scheme of the dipole, shown in Fig. 7.2.

$L_{sc}(t)$, $M_{sc.-n}(t)$], meaning that $\int I(t)\mathrm{d}t$ depends on the variations of the ohmic and inductive parameters of the winding during the quenching process. Longitudinal heat conduction and cooling are neglected, such that T_{hs} will be on the higher, safe side. In practice the MIIT curves are presented in the simplified form

$$T_{hs} = f\left(\int I^2 \mathrm{d}t\right). \tag{7.7}$$

For practical computations of v and T_{hs}, knowing that in a superconductive magnet B varies over the winding length and cross-section, one can subdivide the magnet into about 10 zones of assumed constant B. In addition the peak voltage U_m may eventually be determined from the quench current $I(t)$, the inductive coupling between the normal and superconductive parts of the winding, and of all the resistive parts. U_m may seriously damage the insulation or even destroy the magnet. The highest voltages appear within the excitation coils between pole windings or layers, and not between the main leads, where $\Delta U \approx 0$. Figure 7.1 shows the MIIT curves computed for the 10 m long twin aperture dipole magnet TAP of CERN [7]. Next the influence of the coupling between the pole windings on the quench behaviour will be examined. Figure 7.2 shows the electrical scheme of the TAP magnet. The resulting electrical scheme, including the coupling between windings, is shown in Figure 7.3. The total inductance L [H] of a magnet can be assumed to be

$$L = L_{sc} + L_n + 2M_{sc-n} \tag{7.8}$$

with L_{sc}, L_n the inductances of the superconductive and normal parts of the winding and M_{sc-n} their mutual inductance. The inductances can be obtained from magnetic field computer programs, assuming unit currents flowing in specific parts of the winding. If only a fraction η of a winding is assumed normal, eqn (7.8) becomes

$$L = L_{sc}(1 - \eta^2) + \eta^2 L_n + 2M. \tag{7.9}$$

The resistance of the time dependent normal conducting part of the winding is:

$$R_n = \frac{\rho_n(t)n_t(t)l_t(t)}{A_{Cu}} \tag{7.10}$$

with n_t, l_t the number and average length of the normal conducting turns. The incremental current decay is

$$\Delta I_{m-1}(t) = \frac{-I_{m-1}\Sigma R_n \Delta T}{L} = -I_{m-1}\frac{\Delta t}{\tau} \tag{7.11}$$

$$I_m(t) = I_{m-1} + \Delta I_{m-1}. \tag{7.12}$$

The corresponding voltage drop is

$$\Delta U_m = I_m \Sigma R_n + \frac{\Delta I_{m-1}}{\Delta t}(L_n + M_{sc-n}) \tag{7.13}$$

and the the total extracted energy is

$$\Delta E = \int_0^\infty I_m^2(t) \Sigma R_n(t) dt. \tag{7.14}$$

7.3 External discharge resistor protected and self-protected magnets

The difference between quench protection schemes for individual magnets and for a chain of magnet units, encountered in accelerators and colliders for high energy particle physics, has already been mentioned. Individual magnets can be protected by an external discharge resistor R_d. The total resistance during the quench process is then:

$$R(t) = R_d + R_n(t). \tag{7.15}$$

By using adequate computer codes [3, 4, 5, 8] one has to verify that both the highest temperature in the coils T_{hs} and the maximum voltage U_m remain within permissible limits.

The case of **self-protected magnets** will now be treated. In such a magnet the entire enthalpy of the excitation coils participates in the absorption of the stored electromagnetic energy and the transformation into heat. To this end **quench heaters** are used [10]. These are shown in Fig. 7.4. They consist of thin stainless steel strips, usually insulated by polyimide (kapton) tapes and adequately placed onto and in between the winding layers. The number and location has again to be optimized by computer simulation, aiming at a minimum of heating elements of optimal performance. The heaters are activated by an instantaneous capacitor discharge, triggered by the central quench detector. The delay between the trigger signal and the heater induced quenching can vary in the

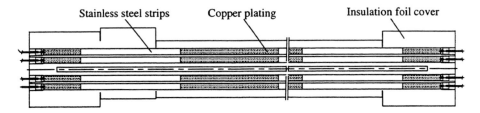

FIG. 7.4. Quench heaters for the LHC magnets (courtesy: CERN, Geneva, Switzerland).

range 20 ms $< \Delta t_h <$ 60 ms. In a chain of superconducting magnets all magnets must be self-protected. However, a second condition must also be satisfied. When a magnet quenches, all magnets of the chain would discharge their energies into the quenching unit and destroy it. Each magnet must thus be protected by a parallel bypassing or short-circuiting diode(s). The diodes are passive elements [11] of diffusion and epitaxial type. Depending upon their location, warm and cold diodes can be distinguished. The first are mounted outside the cooling circuit and require adequate feedthrough. Today one predominantly uses cold diodes, directly placed into the cooling circuit, which simplifies the design and reduces the heat input to the liquid helium. High power diodes will carry maximum currents up to 18 kA, the nominal values amounting to $\approx$ 15 kA. The diodes are characterized by $\int I^2 dt \leq 10^{10} A^2$ or by $w_m = 1[V] \int_0^\infty I dt \leq 1.4$ MJ, where 1 V is the quasi-constant voltage drop of the conducting diode above $T > 50$ K. The inverse voltage of a diffusion-type diode amounts to 60–80 V. Plate 2 shows a test setup of cold diodes for the future LHC collider of CERN.

The maximum voltages, measured on the 15 m long, 8.3 T twin-aperture prototype dipoles for the LHC in the case of a quench, amounted to 2 kV between coils and ground and to 500–600 V between pole windings. The maximum temperature in the coils is determined by the permissible coil elongation due to the temperature gradient. For the LHC dipoles a figure of $T_{hs} \approx 200$ K can be assumed.

The elements of the quench protection system are commanded by a central device, the **quench detector**, consisting of voltage taps, connected at the points of symmetry of a 2n-pole magnet. The pole coils and the matching resistors are connected into equilibrated bridges. In the case of a quench, a voltage signal will appear, triggering other elements of the protection circuit with a delay of $\Delta t_d \approx 20$ ms; the quench heaters are activated after Δt_h and the power switches after $\Delta t_{sw} \approx 25$ ms. One switch will disconnect the power supply feeding the string of magnets, while a second switch will connect the string to an external or dumping resistor R_d.

7.4 Computer codes for quench calculations

As an example the quench simulation code QUABER [4, 5] of CERN will be discussed. The code solves the set of equations (7.1)–(7.14). The working chart is shown in Fig.

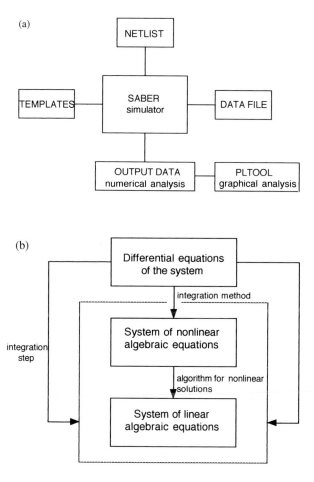

FIG. 7.5. (a) Working chart of the QUABER program. (b) Working chart of the analogue
    SABER code.

7.5a. The program input contains four groups of information: NETLIST contains the information related to the electrical circuit and connections, DATAFILE the information
on the cable and coil geometries, the magnet load lines for every section of the assumed
constant magnetic induction B, and the heater location; TEMPLATES are sub-programs
which describe the electrical and thermal characteristics of each element (quench resistance, inductance, couplings and related voltages), and compute the non-linear elements
such as $\rho(B, T)$, $C(T)$, and the MIIT curves. Figure 7.5b shows how the analogue code
SABER solves the differential equations which determine the quench simulation process: SABER can handle a large number of parameters of the non-linear behaviour. It
contains three integration methods to transform a set of differential equations into non-

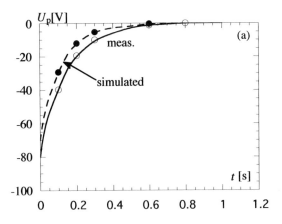

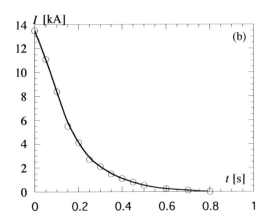

FIG. 7.6. (a) QUABER-code simulated and measured pole voltage during a quench in an LHC twin aperture dipole model of 1 m length. (b) QUABER-code simulated and measured decay current during a quench in the same magnet.

linear algebraic equations by the methods of Newton–Raphson and Katznelson.

The QUABER program can perform general electromagnetic and thermal calculations. Applied to quench calculations it will yield the voltages between layers and pole windings, the inner and outer extracted energy, the transverse turn-to-turn quench spreading, compute the normal zone length, and the effect of independent firing of several quench heaters. As a special feature the program will compute the quench back-firing effect, due to excessive $\frac{dI}{dt}$ or to currents exceeding $I_c(B)$. A graphics package allows the presentation of the thermal and electric parameters as time functions at any

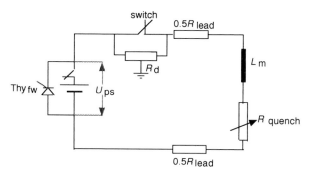

FIG. 7.7. Simplified electrical scheme for quench tests on the 10 m long LHC prototype
dipole.

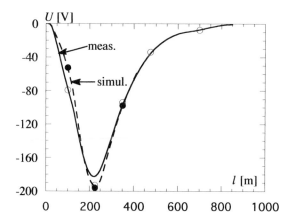

FIG. 7.8. Measured and QUABER-simulated outer layer voltage after a quench in a
10 m LHC dipole without external energy extraction.

point of the excitation coils giving an answer to the question of whether a magnet is self-
protected or if quench heaters—where, how many—are required. As an example for the
QUABER calculations, the measured and computed pole voltage and the current decay
in a 1 m long, twin aperture dipole magnet model for the LHC are shown in Figs 7.6a
and 7.6b. More recent test results and comparative simulations related to the first 10 m
long prototype dipole for the LHC are shown in Figs 7.7 and 7.8 [12, 13]. The resulting
electrical scheme is shown in Fig. 7.7, reproducing closely the conditions of a dipole
mounted in a string of LHC magnets. To simulate the short-circuiting power diode,
the free-wheeling parallel thyristors Thy$_{fw}$ of the power supply are activated, once the

supply is switched off. During that operation the switch across the dumping resistor R_d remains closed. Figure 7.8 shows the measured and QUABER-simulated voltage U_1 across an outer coil layer of the dipole depending on the winding length.

To conclude it should be stated that the QUABER program can only yield correct results if the input parameters are correct. Whenever necessary, they should be determined by adequate experiments.

TRANSVERSE BEAM DYNAMICS IN CIRCULAR PARTICLE ACCELERATORS FOR HIGH ENERGY PHYSICS

8.1 Introduction

Classical and superconducting magnets and conventional and superconductive high frequency cavities are the basic elements in any circular accelerator and collider. The magnets determine the particle or beam behaviour in the **transverse** plane, the cavities in the **longitudinal**. Specific beam oscillations, related to the interplay of both transverse and longitudinal phenomena, will not be considered.

Since we are mainly interested in the quality and requirements on superconducting accelerator magnets, a short introduction to transverse beam dynamics and beam optics will be given. The success of any accelerator or collider for high energy physics will, among others, depend on a close harmonization and mutual matching between transverse beam dynamics requirements and the superconducting magnet performances from the design stage to the completion and operation of the accelerator.

8.2 Particle motion in a fixed and curved coordinate system

For the guiding of charged particle beams one needs **bending, focusing,** and **correcting** magnets. Assuming a fixed x, z, s coordinate system, shown in Fig. 8.1, the Lorentz force, acting on a charged particle moving in the longitudinal direction with the velocity v [m s^{-1}], exposed to a vertical magnetic induction B_z [T] and a transverse electrical field E_x [V m^{-1}], is equal to [1, 2, 3, 4]:

$$F_x = \frac{\mathrm{d}\,\vec{p}}{\mathrm{d}x} = \frac{\mathrm{d}}{\mathrm{d}t}(m\,\vec{v}) = e[\vec{E}_x + \vec{v} \times \vec{B}_z]. \tag{8.1}$$

The advantage of $\vec{B}$ over $\vec{E}$ in eqn (8.1) can be easily demonstrated. Assuming that the particle velocity is $v \approx c$ and close to 3×10^8 [m s^{-1}], one obtains at a rather low induction of $B = 1$ T, $vB = 3 \times 10^8$ [V m^{-1}]; the high equivalent electrical field would amount to $E = 3 \times 10^8$ [V m^{-1}]. In what follows only transverse magnetic fields will be considered. In a constant and homogeneous vertical field B_z with $B_x = B_s = 0$ one obtains from equating the Lorentz and the centrifugal forces:

$$ev B_z = -\frac{mv^2}{\rho}; \quad -\frac{1}{\rho} = \frac{x''}{(1 + x'^2)^{1.5}} = \frac{eB_z}{p}. \tag{8.2}$$

The particle moves on a circle of radius

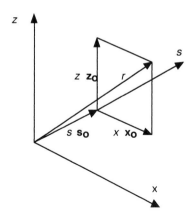

FIG. 8.1. Fixed x, z, s coordinate system.

$$-\frac{1}{\rho}[\mathrm{m}] = 0.3\frac{B_z[\mathrm{T}]}{p[\mathrm{GeV\,c^{-1}}]}.\tag{8.3}$$

In the x, z, s coordinate system, s is the curved trajectory of the central particle or reference trajectory. For small excursions of s, according to Fig. 8.2 a cylindrical coordinate system z, r, θ can be chosen, with $r = p + x$ and $\theta = \frac{s}{\rho}$. For particle motion in the horizontal midplane $z = 0$ and for small excursions in the x-direction, stability is obtained for

$$eB_z(r)\begin{array}{c}<\\>\end{array}\frac{mv}{r} \quad \text{for } r\begin{array}{c}<\\>\end{array}\rho.\tag{8.4}$$

Introducing the field variation index

$$n = \frac{\rho}{B_0}\bullet\left(\frac{\partial B_z}{\partial r}\right)_{r=\rho}\tag{8.5}$$

where B_0 is the field at the centre, eqn (8.2) becomes

$$evB_z(r) \approx ecB_0\left(1 - n\frac{x}{\rho}\right)_{r=\rho}\tag{8.6}$$

and

$$\frac{mv^2}{r} = \frac{mv^2}{\rho}\left(1 - n\frac{x}{\rho}\right)\tag{8.7}$$

According to eqn (8.4) stability is obtained for

$$0 < n < 1.\tag{8.8}$$

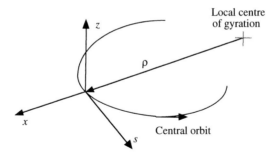

FIG. 8.2. Curved coordinate system following a reference trajectory.

For small excursions in x and z,

$$x'' = \frac{\partial^2 x}{\partial s^2}; \ z'' = \frac{\partial^2 z}{\partial s^2}$$

which can be written

$$x'' + \omega_0^2(1 - n)x = 0 \tag{8.9}$$

$$z'' + \omega_0^2 nz = 0 \tag{8.10}$$

where $\omega_0 = \frac{eB_0}{m}$ is the revolution or cyclotron frequency of the particle:

$$f_x = \sqrt{1 - n}\, f_0 = \sqrt{1 - n}\frac{\omega_0}{2\pi} \tag{8.11}$$

$$f_z = \sqrt{n}\, f_0 = \sqrt{n}\frac{\omega_0}{2\pi}. \tag{8.12}$$

Equations (8.9) and (8.10) describe the particle motion in a **weak focusing** accelerator. As $0 < n < 1$, the oscillation frequencies in the horizontal and vertical planes are lower than the revolution frequency f_0 and less than one full betatron oscillation per turn. To satisfy condition (8.8) a magnet profile according to Fig. 8.3 is required such that

$$\frac{\partial B_z}{\partial r} = \frac{\partial B_x}{\partial z} < 0. \tag{8.13}$$

The drawback of the now historical weak focusing machines was the required large aperture which implied the use of large and heavy magnets of relatively modest inductions.

This limitation was overcome by the proposal of the **strong focusing** principle in 1952 by Courant, Snyder and Livingstone [5] and independently also by Christophilos in 1951 [6]. It triggered the development of accelerators with alternating magnet sectors

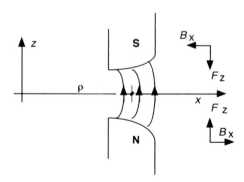

FIG. 8.3. Weak focusing synchrotron magnet.

of strongly increasing magnetic inductions with radius $\frac{\partial B_z}{\partial r} \gg 1$ or $n \ll -1$ and of strongly decreasing ones with $n \gg +1$, as shown in Fig. 8.4. In analogy to a pair or doublet of focusing and defocusing optical lenses such a magnet system will under certain conditions be focusing in both the vertical and horizontal plane. Equations (8.9) and (8.10) are special cases of the general linear trajectory equations in the x, z, s coordinate system:

$$z'' + kz = 0 \qquad (8.14)$$

$$x'' - \left(k - \frac{1}{p^2} \right) = \frac{1}{\rho} \frac{\delta p}{p_0} \qquad (8.15)$$

where

$$p = p_0 \left(1 + \frac{\Delta p}{p_0} \right). \qquad (8.16)$$

is the off-equilibrium orbital momentum which can also be written as

$$\frac{1}{p} \approx \frac{1}{p_0} \left(1 - \frac{\Delta p}{p_0} \right). \qquad (8.17)$$

In the general case where the bending strength $\frac{1}{\rho(s)}$ [m^{-1}] and the quadrupole strength $k(s) = \frac{e}{p} \frac{\partial B_z}{\partial x}$ [m^{-2}] vary along the orbit, eqns (8.14) and (8.15) become Hill-type equations:

$$y'' + K(s)y = \frac{1}{\rho(s)} \frac{\Delta p}{p_0}. \qquad (8.18)$$

The general solutions to this equation are

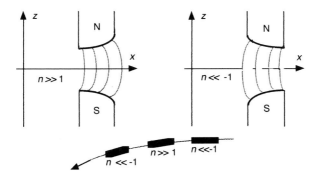

FIG. 8.4. Alternating gradient focusing principle.

$$y(s) = C(s)y_0 + S(s)y_0' + D(s)\frac{\Delta p}{p_0} \tag{8.19a}$$

$$y_0'(s) = C'(s)y_0 + S'(s)y_0' + D'(s)\frac{\Delta p}{p_0} \tag{8.19b}$$

where $C(s)$ and $S(s)$, the cos-like and sin-like functions, are the two independent solutions to the homogeneous Hill-equation (for $\frac{1}{\rho}\frac{\Delta p}{p_0} = 0$) satisfying the initial condition

$$\begin{pmatrix} C_0 & S_0 \\ C_0' & S_0' \end{pmatrix} = \begin{pmatrix} 1 & 0 \\ 0 & 1 \end{pmatrix}. \tag{8.20}$$

The dispersion D is a particular solution to the inhomogeneous Hill-equation for $\frac{\Delta p}{p_0}$ with the initial conditions

$$\begin{pmatrix} D_0 \\ D_0' \end{pmatrix} = \begin{pmatrix} 0 \\ 0 \end{pmatrix}. \tag{8.21}$$

In matrix notation the general solution to the Hill-equation can be written

$$\begin{pmatrix} y \\ y' \end{pmatrix}_s = \begin{pmatrix} C & S \\ C' & S' \end{pmatrix} \begin{pmatrix} y \\ y' \end{pmatrix}_0 + \frac{\Delta p}{p_0} \begin{pmatrix} D \\ D' \end{pmatrix} \tag{8.22a}$$

$$\begin{pmatrix} y \\ y' \\ \frac{\Delta p}{p_0} \end{pmatrix}_s = \begin{pmatrix} C & S & D \\ C' & S' & D' \\ 0 & 0 & 1 \end{pmatrix} \begin{pmatrix} y \\ y' \\ \frac{\Delta p}{p_0} \end{pmatrix}_0. \tag{8.22b}$$

Since the initial matrix is equal to 1 and the subsequent matrices do not depend upon s the total transformation matrix remains equal to 1 throughout the system. **The stability criterion** for particles of central momentum p_0 can be determined as follows: the transfer matrix for a period L of the accelerator structure is

$$M(s) = M\left(s + \frac{L}{s}\right), \tag{8.23}$$

for a full revolution containing NL periods,

$$M\left(s + \frac{NL}{s}\right) = M(s)^N \tag{8.24}$$

and for n turns

$$M\left(s + \frac{nNL}{s}\right) = [M(s)]^{nN}. \tag{8.25}$$

Stable motion is obtained when the elements of the matrix according to eqn (8.25) remain bounded for $n \to \infty$. The condition to be satisfied is that the eigenvalues of M satisfy the relation

$$MY = \lambda I; \quad Y = \begin{pmatrix} y \\ y' \end{pmatrix}; \quad I = \begin{pmatrix} 1 & 0 \\ 0 & 1 \end{pmatrix}. \tag{8.26}$$

Expressing the matrix $M = \begin{pmatrix} a & b \\ c & d \end{pmatrix}$, one obtains:

$$\lambda^2 - \lambda(a + d) + ad - bc = 0 \tag{8.27a}$$

$$\lambda_{1,2} = \frac{a+d}{2} \pm j\sqrt{1 - \frac{(a+d)^2}{4}} = \cos \mu + j \sin \mu \tag{8.27b}$$

and

$$\cos \mu = \frac{1}{2}\mathrm{trace} M = \frac{1}{2}(a + d). \tag{8.28}$$

M can also be expressed by the Twiss matrix

$$M = I \cos \mu + J \sin \mu = \begin{pmatrix} 1 & 0 \\ 0 & 1 \end{pmatrix} \cos \mu + \begin{pmatrix} \alpha & \beta \\ -\gamma & -\alpha \end{pmatrix} \sin \mu \tag{8.29}$$

where

$$\alpha = \frac{a - d}{2 \sin \mu}; \quad \beta = \frac{b}{\sin \mu}; \quad \gamma = -\frac{c}{\sin \mu}. \tag{8.30}$$

For M^n one obtains

$$M^n = I \cos(n\mu) + J \sin(n\mu). \tag{8.31}$$

When $n \to \infty$ M^n will be bounded and stability reached for

$$|\text{Trace}\,M| < 2; \mu \to \text{real}. \tag{8.32}$$

The Twiss parameters α, β, γ are linked by the following relations

$$\alpha(s) = -\frac{1}{2}\beta'(s) \tag{8.33}$$

$$\gamma(s) = \frac{1 + \alpha^2(s)}{\beta(s)} = \frac{1 + \frac{1}{4}\beta'^2(s)}{\beta(s)}. \tag{8.34}$$

Hill's equation can now be expressed by

$$y_{1,2}(s) = a\sqrt{\beta(s)}\,\exp[\pm\Phi(s)] \tag{8.35}$$

with a the amplitude and λ the wavelength of the pseudo-harmonic oscillation:

$$\lambda(s) = 2\pi\beta(s). \tag{8.36}$$

The phase function is:

$$\Phi(s) = \int_{s_0}^{s} \frac{dt}{\beta(s)}; \, \Phi' = \frac{1}{\beta(s)} \tag{8.37}$$

and the phase advance per cell over the period L is

$$\mu = \int_{s}^{s+L} \frac{dt}{\beta(s)} = \Phi(s + L, s). \tag{8.38}$$

The Q (or ν) value of an accelerator, defined as the number of betatron oscillations per revolution, is equal to

$$Q = \frac{N\mu}{2\pi} = \frac{1}{2\pi} \oint \frac{ds}{\beta(s)}. \tag{8.39}$$

A particle trajectory is described by a real solution of the Hill equation:

$$y(s) = a\sqrt{\beta(s)}\cos[\Phi(s) - \delta] \tag{8.40a}$$

$$y'(s) = -\frac{a}{\sqrt{\beta(s)}}\left\{-\frac{1}{2}\beta'\cos[\Phi(s) - \delta] + \sin[\Phi(s) - \delta]\right\} \tag{8.40b}$$

with the arbitrary phase constant δ. For a fixed value of s a family of trajectories, having the same amplitude but different phases δ, can be represented by an ellipse in $Y(y, y')$-phase space, as shown in Fig. 8.5. For $\delta = 2\pi$ a full turn around the ellipse contour is made. The ellipse area is given by:

$$S = a^2\pi = \varepsilon\pi = \pi(\gamma y^2 + 2\alpha yy' + \beta y'^2). \tag{8.41}$$

The right-hand expression is the **Snyder–Courant invariant**, which remains constant for a particle trajectory within an accelerator. If now a particle beam centred around

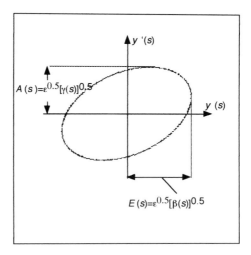

FIG. 8.5. The phase space ellipse.

the origin fills in point s the phase ellipse up to the amplitude $E(s)$, then all trajectories will remain within the ellipse when moving within the accelerator. Since the ellipse area $a^2\pi$ or $\varepsilon\pi$ is constant, the phase-space density is also an invariant. This is the important **theorem of Liouville**, valid for a constant particle momentum p and excluding stochastic, beam scattering, synchrotron radiation, or non-linear correction effects. The beam envelope, defined by $E(s)$ and by the maximum beam divergence $A(s)$:

$$E(s) = y_m(s) = \sqrt{\varepsilon}\sqrt{\beta(s)} \tag{8.42}$$

$$A(s) = y'_m(s) = \sqrt{\varepsilon}\sqrt{\gamma(s)} \tag{8.43}$$

is shown in Fig. 8.5.

Particles with a momentum deviation of $\Delta p = p - p_0$ will satisfy Hill's equation in the horizontal plane, according to eqns (8.15) and (8.17). The total deviation from the reference orbit will be:

$$x(s) = x_D(s) + x_\beta(s) = D(s)\frac{\Delta p}{p_0} + x_\beta(s). \tag{8.44}$$

The periodic dispersion satisfies the differential equation:

$$D''(s) + K(s)D(s) = \frac{1}{\rho(s)} \tag{8.45}$$

and $x_D(s)$ is the deviation from the reference orbit of the off-momentum closed orbit and $x_\beta(s)$ stands for the betatron oscillation around the dispersion orbit. Imposing periodic boundary conditions $D(s + L) = D(s + NL) = D(s)$ and assuming that the vertical plane also contains a perturbing term $F(s)$ due to field and alignment errors in the magnet, the closed orbit periodic dispersion is obtained for $F(t) = \frac{1}{\rho(t)}$ as

$$D(s) = \frac{\sqrt{\beta(s)}}{2\sin(\pi Q)} \oint \frac{\sqrt{\beta(t)}}{\rho(t)} \cos[\Phi(t) - \Phi(s) - \pi Q] \, dt. \tag{8.46}$$

Equation (8.46) states that finite dispersion exists only for a number of betatron oscillations per revolution Q different from an integer. The relative closed orbit variation with the relative momentum deviation, known as the momentum compaction factor, is also related to the dispersion $D(s)$:

$$\alpha_m = \frac{\frac{\Delta L}{L}}{\frac{\Delta p}{p_0}} = \frac{1}{L} \int \frac{D(s)}{\rho(s)} \, ds. \tag{8.47}$$

8.3 Accelerator magnets and their transfer matrices

The main accelerator magnets and their transfer matrices will now be discussed. There exist several types of **bending magnets** for accelerators, providing a uniform vertical induction $B_z = B_0$. According to eqn (8.3) the radius of curvature $\frac{1}{\rho}$ for a particle of charge e, moving in B_0 is $0.2998 \frac{B_0}{p}$ [Tc/(GeV)]. In high energy physics accelerators with large values of ρ and individual magnet length $l \ll \rho$ one can assume that the beam central trajectory is perpendicular to the magnet faces. In that case $\frac{1}{\rho} \approx 0$, the transfer matrix will be equal to that of the **drift space** without any focusing term, $k = 0$, and $\frac{1}{\rho} = 0$. For the dipole sector magnet of length l shown in Fig. 8.6 and a central beam trajectory perpendicular to the end faces one obtains:

$$\varphi = \frac{l}{\rho}. \tag{8.48}$$

A rectangular magnet with ends which are not perpendicular to the central beam trajectory—see Fig 8.7—can be considered as a sector magnet with two wedges of a deflection angle

$$\alpha = \frac{\Delta l}{\rho} = \frac{x \operatorname{tg} \delta}{\rho} = \frac{x}{f}. \tag{8.49}$$

The wedges act as defocusing lenses in the horizontal plane and focusing lenses in the vertical plane. The net focusing effect in the horizontal plane is zero, as the defocusing effect of the ends is compensated by the focusing of the sector magnet part; in the vertical plane the magnet will be weakly focusing.

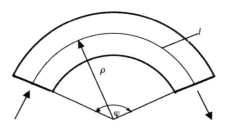

FIG. 8.6. The dipole sector magnet.

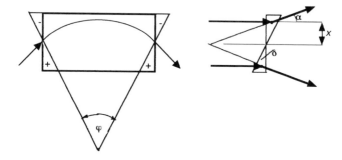

FIG. 8.7. The rectangular dipole magnet.

A **quadrupole magnet**, shown in Fig. 8.8, is characterized by the linear field components

$$B_z = -gx; \quad B_x = -gz \tag{8.50}$$

of **constant field gradient** g [Tm^{-1}]

$$g = -\frac{dB_z}{dx} = -\frac{dB_x}{dz}. \tag{8.51}$$

In the central conductor- and iron-free part of a quadrupole the magnetic induction and the gradient can be derived from the scalar potential $V(x, z)$ [Tm]:

$$V(x, z) = gxz. \tag{8.52}$$

From eqn (8.50) one obtains

$$|B| = (B_x^2 + B_z^2)^{0.5} = [(gx)^2 + (gz)^2]^{0.5} = gr. \tag{8.53}$$

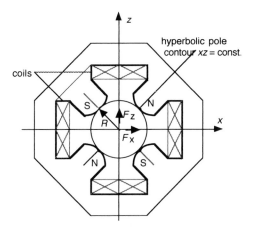

FIG. 8.8. The quadrupole lens.

In analogy to the bending strength $\frac{1}{\rho}$ of a dipole, the quadrupole strength

$$k_q = \frac{eg}{p_0} \tag{8.54}$$

can be defined. Figure 8.9 shows the important thin quadrupole doublet, composed of two quadrupoles of small physical length compared to their focal length $l_q \ll f$ and

$$k_q l_q \approx \frac{1}{f}. \tag{8.55}$$

A focusing and a defocusing quadrupole lens having the same strength $\frac{1}{f_1} = -\frac{1}{f_2} = \frac{1}{f}$, separated by a drift space of length l, will give a doublet of strength:

$$\frac{1}{f} = \frac{1}{f_1} + \frac{1}{f_2} - \frac{l}{f_1 f_2} = \frac{l}{f^2}. \tag{8.56}$$

The doublet will have the same focusing effect in the two planes, whereby the drift space will not exceed the double focal length $l < 2f$.

The **sextupole magnet**, shown in Fig. 8.10, is important for correcting chromatic errors in accelerators and non-linear field errors in bending and quadrupole magnets. The magnetic field components of a sextupole are

$$B_z = \frac{1}{2} g'(x^2 - z^2) \tag{8.57}$$

$$B_x = g'xz \tag{8.58}$$

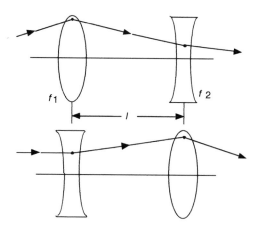

FIG. 8.9. Thin lens doublet.

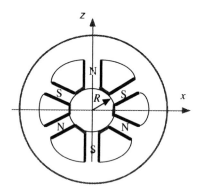

FIG. 8.10. The sextupole lens.

which can be derived from the scalar potential

$$V(x, z) = -\frac{1}{2}g'\left(x^2 z - \frac{1}{3}z^3\right)$$

(8.59)

where

$$B_R = \sqrt{B_z^2 + B_x^2} = -\frac{1}{2}g'R^2$$

(8.60)

and

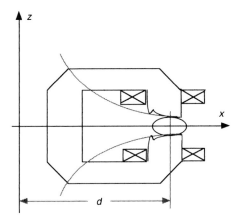

FIG. 8.11. The combined function synchrotron magnet.

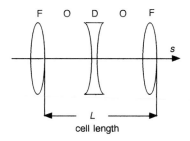

FIG. 8.12. The FODO magnet structure.

$$g' = \frac{d^2 B_R}{dR^2}. \tag{8.61}$$

The combined function synchrotron magnet, shown in Fig. 8.11, has a bending and quadrupole component. The magnetic field can be derived from the scalar potential

$$V(x, z) = -B_0 z + gxz \tag{8.62a}$$

$$B_x = gz \tag{8.62b}$$

$$B_z = -B_0 + gx. \tag{8.62c}$$

The particle trajectories in any of the above magnets of length l between entrance and exit, defined by the indices 1 and 2 with the exception of the non-linear sextupole, are given by the following matrix relations:

For the focusing case $K > 0$

$$\left\{ \begin{matrix} y \\ y' \end{matrix} \right\}_2 = \left\{ \begin{matrix} \cos(|K|^{0.5}l) & |K|^{-0.5} \bullet \sin(|K|^{0.5}l) \\ -|K|^{0.5} \bullet \sin(|K|^{0.5}l) & \cos(|K|^{0.5}l) \end{matrix} \right\} \left\{ \begin{matrix} y \\ y' \end{matrix} \right\}_1 ; \qquad (8.63)$$

for the defocusing case $K > 0$

$$\left\{ \begin{matrix} y \\ y' \end{matrix} \right\}_2 = \left\{ \begin{matrix} \cosh(|K|^{0.5}l)|K|^{-0.5} \bullet \sinh(|K|^{0.5}l) \\ |K|^{0.5} \bullet \sinh(|K|^{0.5}L) & \cosh(|K|^{0.5}l) \end{matrix} \right\} \left\{ \begin{matrix} y \\ y' \end{matrix} \right\}_1 ; \qquad (8.64)$$

and for the drift space case, $K = 0$

$$\left\{ \begin{matrix} y \\ y' \end{matrix} \right\}_2 = \left\{ \begin{matrix} 1 & l \\ 0 & 1 \end{matrix} \right\} \left\{ \begin{matrix} y \\ y' \end{matrix} \right\}_1 . \qquad (8.65)$$

For the dispersion D one obtains for
$K > 0$

$$\left\{ \begin{matrix} D \\ D' \end{matrix} \right\} = \left\{ \begin{matrix} \dfrac{1}{\rho|K|} \bullet (1 - \cos\varphi) \\ \dfrac{1}{\rho\sqrt{|K|}} \bullet \sin\varphi \end{matrix} \right\} ; \qquad (8.66a)$$

for $K < 0$

$$\left\{ \begin{matrix} D \\ D' \end{matrix} \right\} = \left\{ \begin{matrix} -\dfrac{1}{\rho|K|} \bullet (1 - \cosh\varphi) \\ \dfrac{1}{\rho\sqrt{|K|}} \bullet \sinh\varphi \end{matrix} \right\} ; \qquad (8.66b)$$

and for $|K| = 0$

$$\left\{ \begin{matrix} D \\ D' \end{matrix} \right\} = \left\{ \begin{matrix} 0 \\ 0 \end{matrix} \right\} , \qquad (8.66c)$$

with $\varphi = \dfrac{l}{\rho} = l\sqrt{|K|}$.

The parameters shown in Table 8.1 have to be introduced for K_x and K_z for different types of accelerator magnets:

As an example the transfer matrix of the magnet structure of the superconducting collider HERA will be computed. The $L = 47$ m long cell has a FODO structure with one focusing and one defocusing quadrupole of $l_q = 1.9$ m, and with two 10 m long bending magnets, assumed to be drift spaces since $\frac{l_b}{\rho_x} = \frac{10}{584} \cong 0$. The quadrupoles can be treated as thin lenses. The transfer matrix for a cell is:

$$M = \{M_F M_O M_D M_O\} = \left\{ \begin{matrix} 1 + \dfrac{L}{2f} & L + \dfrac{L^2}{4f} \\ -\dfrac{L}{2f^2} & 1 - \dfrac{1}{2f} - \dfrac{L^2}{4f^2} \end{matrix} \right\} . \qquad (8.67)$$

Table 8.1 *Parameters for K_x and K_z*

Type of magnet	$K_x[\mathrm{m}^{-2}]$	$K_z[\mathrm{m}^{-2}]$	Fig. No.
Combined function magnet	$\rho_x^{-2} - k$	k	8.11
Vertical dipole magnet	ρ_x^{-2}	0	8.6
Rectangular dipole magnet	ρ^{-2}	ρ^{-2}	8.7
Quadrupole magnet	$-k$	k	8.8
Thin quadrupole lens	$\frac{1}{fl_q}$	$-\frac{1}{fl_q}$	8.9
Drift space or vertical dipole magnet with $\rho_x \approx \infty$	0	0	

Comparing the result with the Twiss matrix representation according to eqn (8.29) one finds

$$\cos \mu = \frac{1}{2}\mathrm{trace}M = 1 - \frac{L^2}{8f^2}; \; |\sin\frac{\mu}{2}| = \frac{L}{4f}. \tag{8.68}$$

The stability criterion requires $\frac{L}{4f} < 1$; $f > \frac{L}{4}$. For a 90° phase advance $-\cos\mu = 0 - f = \frac{1}{2\sqrt{2}}L$. The HERA magnet system meets these conditions since $L = 47$ m, $k = 0.033$ m^{-2}, $l_q = 1.9$ m, $f = 16.4$ m, and $\mu \cong 90°$. In a similar way further machine parameters like β, D, D' can be computed [1].

So far only ideal accelerator magnet configurations have been considered: pure dipole, quadrupole, and sextupole magnets with rigorously constant, linear, and sextupolar fields. Ideal magnets do not exist in practice, and any magnet, classical or superconductive, will exhibit **field errors**. They can be decomposed into direct or in phase and skew or turned field errors and expressed in terms of higher order (sometimes also in terms of lower order) direct and skew harmonics. Most of these errors are of geometrical, manufacturing, and of electromagnetic origin. Geometrical errors are caused by mechanical tolerances, positioning errors of the excitation coils and conductors, while the electromagnetic errors are related to the remanent iron magnetization, iron saturation, the magnetization currents in the superconductors, and to eddy current effects. Any magnetic field error, any deviation from the assumed theoretical value will act as a perturbing term in the Hill equation and increase the beam dispersion or blow-up. Special attention must be paid to field harmonics which may excite stop-band resonances [7]. Magnetic field errors will be treated later; their impact and consequences for the beam dynamics are treated in references [7, 8].

9

WINDING CONFIGURATIONS FOR SUPERCONDUCTING ACCELERATOR MAGNETS

9.1 Introduction

In the preceding chapter on transverse beam dynamics examples of classical 2n-pole magnets for particle accelerators have been given. The steady development and improvement of technical superconductors of high j_c and B as described in Chapter 5, led to a vigorous development of superconducting magnets for accelerators and colliders; an energy increase by a factor of 4–5 could be achieved solely by increasing B. A number of successfully operating accelerators and colliders like the 1 GeV Tevatron in the USA and the 800 GeV HERA in Germany were built and the design and construction of the ambitious project of the 7 TeV–7 TeV hadron collider LHC of CERN is under way.

There are several compelling reasons to exclude classical magnet configurations for these machines:

(i) In classical magnets it is the iron contour or pole shape which determines the field quality. Due to early iron saturation at $1.5T < B < 2T$ an excitation of iron-dominated superconducting magnet windings would rapidly result in considerable field distortions. Earlier efforts in this direction such as the proposal to use so-called 'superferric' dipoles for the regretfully abandoned 20 TeV–20 TeV SSC collider had not been retained [1].

(ii) Classical magnet configurations are not suited for optimal exploitation of superconducting NbTi or Nb_3Sn cables in the 5–20 kA range, notably for quadrupoles.

(iii) The containment of mechanical forces and their transmission via the cryostat to the structure at ambient temperature would be complicated and costly.

It was soon perceived that geometrically simple and straightforward configurations of excitation coils, force retaining system, iron screen and cryostat will considerably increase the performance of superconducting accelerator magnets. Today current-carrying conductor configurations, generating the required field, are generally used, whereby the cylindrical, coaxial, and concentric design of circular aperture, coils, mechanical system and iron screen is commonly applied. Recent investigations about this configuration, commonly known as the '$\cos \theta$' design, have shown that it could be used for superconducting magnets up to 15 T. An iron screen is usually placed at a radius where possibly $B \leq 2T$, but even at higher fields an iron contribution of 10–20% to the total bore field can be expected.

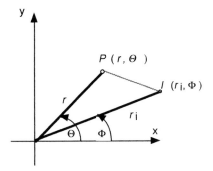

FIG. 9.1. Vector potential at point P of a single current-carrying wire.

9.2 Magnetic field calculations for configurations of current carrying superconductors

Assuming magnets with a large length-to-width ratio which is equivalent to neglecting end effects, the magnetic field in the transverse, complex z-plane

$$z = x + iy \tag{9.1}$$

can be derived from the longitudinal vector potential $A_s[\mathrm{Vsm}^{-1}]$ [2]:

$$B_y = \frac{\partial A_s}{\partial x} = \frac{\partial}{\partial x}\left[\sum_n A_n f_n(x, y)\right] = \sum_{n=1}^{\infty} \frac{x^{n-1}}{(n-1)!} \cdot \frac{\partial^{n-1} B_y}{\partial x^{n-1}} \tag{9.2}$$

so that

$$A_n = \frac{1}{n!}\left(\frac{\partial^{n-1} B_y}{\partial x^{n-1}}\right)_{y=0} . \tag{9.3}$$

In a polar coordinate system (s, r, θ) one obtains for A_s at a point $P(r, \theta)$, of a current carrying wire $I(r_i, \Phi)$, according to Fig. 9.1 [3]:

$$A_s = \frac{\mu_0 I}{2\pi} \sum \frac{1}{n}\left(\frac{r}{r_i}\right)^n \cos[n(\Phi - \Theta)]. \tag{9.4}$$

The magnet induction components B_Θ, B_Φ are

$$B_\Theta = -\frac{\partial A_s}{\partial r} = -\sum_{n=1}^{\infty} \frac{\mu_0 I}{2\pi r_i} \left(\frac{r}{r_0}\right)^{n-1} \cos[n(\Phi - \Theta)] \qquad (9.5)$$

$$B_r = \frac{1}{r}\frac{\partial A_s}{\partial \Theta} = -\sum_{n=1}^{\infty} \frac{\mu_0 I}{2\pi r_i} \left(\frac{r}{r_0}\right)^{n-1} \sin[n(\Phi - \Theta)] \qquad (9.6)$$

When computing B_Θ, B_Φ for several current carrying wires, on has to sum up all the individual contributions. The relations between B_x, B_y and B_r, B_Θ are

$$B_x = B_r \cos\Theta - B_\Theta \sin\Theta \qquad (9.7)$$

$$B_y = B_r \sin\Theta + B_\Theta \cos\Theta. \qquad (9.8)$$

B_r and B_Θ are conveniently expressed by a Fourier multipole expansion around a reference radius r_0 and the main field harmonic B_0:

$$B_\Theta = B_0 \sum_{n=1}^{\infty} \left(\frac{r}{r_0}\right)^{n-1} (b_n \cos n\Theta + a_n \sin n\Theta) \qquad (9.9)$$

$$B_r = B_0 \sum_{n=1}^{\infty} \left(\frac{r}{r_0}\right)^{n-1} (-a_n \cos n\Theta + b_n \sin n\Theta) \qquad (9.10)$$

where b_n and a_n are the **normal** and **skew** multipole coefficients and n is the harmonic number. For multipoles symmetric with respect to the horizontal axis:

$$n = (1 + 2k)m \qquad (9.11)$$

with $m = 1, 2, 3, \ldots$ and $k = 0, 1, 2, 3, \ldots$. For a dipole ($m = 1$) the harmonics are $n = 1, 3, 5, \ldots$, and for a quadrupole $n = 2, 4, 6, \ldots$. The basic harmonic is equal to the number of pole pairs $n = p = 1$ for a dipole, $n = p = 2$ for a quadrupole, etc. Equations (9.9) and (9.10) can also be derived from the magnetic scalar potential

$$V(r) = V_0 + \sum_{n=1}^{\infty} \left[a_n \left(\frac{r}{r_0}\right)^n \sin n\Theta + b_n \left(\frac{r}{r_0}\right)^n \cos n\Theta\right]. \qquad (9.12)$$

The vertical field component on the median plane at a distance $x = r/r_0$ is obtained for $\Theta = 0$, $B_\Theta = B_y$:

$$B_y = b_1 + 2b_2 x + 3b_3 x^2 + \ldots + n b_n x^{n-1} \tag{9.13}$$

where b_n are measured at r_0. The **field error** of the harmonic n at x is equal to:

$$\frac{B_n}{B_1(0)} = \frac{n b_n x^{n-1}}{b_1} \tag{9.14}$$

or

$$\frac{\partial^{n-1} B_n}{\partial x^{n-1}} = n! b_n. \tag{9.15}$$

In quadrupoles one defines the gradient errors:

$$\frac{g_n(x)}{g_2(0)} = \frac{\dfrac{\partial B_n(x)}{\partial x}}{\dfrac{\partial B_2(0)}{\partial x}} = \frac{n(n-1)b_n x^{n-2}}{2b_2} \tag{9.16}$$

Equation (9.13) corresponds to the European notation. The American notation is:

$$B_y = B_0(1 + b_1 x + b_2 x^2 + \ldots + b_n x^n). \tag{9.17}$$

In the general case skew multipole coefficients a_n will also appear and will have to be added into eqns (9.13) and (9. 17).

In order to calculate adequate and realistic conductor configurations yielding dipole, quadrupole, and higher multipole fields in a circular aperture of radius ρ, it shall be proved that a line current density pattern of $I_1 [\mathrm{Am}^{-1}]$ in accordance with Fig. 9.2

$$I_1(\Phi) = j_{10} R \cos p\Phi = I_0 \cos p\Phi \tag{9.18}$$

would generate such multipole fields. In fact, integrating eqn (9.4) one obtains:

$$A_s(r, \Theta) = \frac{\mu_0 I_0}{2p} \left(\frac{r}{R}\right)^p \cos p\Theta \tag{9.19}$$

and, for B_Θ and B_r,

$$\left\{ \begin{matrix} B_\Theta \\ B_r \end{matrix} \right\} = -\frac{\mu_0 I_0}{2R} \left(\frac{r}{R}\right)^{p-1} \left\{ \begin{matrix} \cos p\Theta \\ \sin p\Theta \end{matrix} \right\}. \tag{9.20}$$

With eqns (9.7) and (9.8) one obtains for the dipole ($p = 1$)

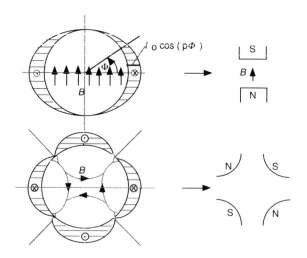

FIG. 9.2. $\cos p\Phi$ line current repartition around a circular aperture yielding ideal multipole fields.

$$B_x = 0; \quad B_y = -\frac{\mu_0 I_0}{2R} \tag{9.21}$$

and for the quadrupole

$$B_x = gy = -\frac{\mu_0 I_0}{2R^2}y \tag{9.22}$$

$$B_y = gx = -\frac{\mu_0 I_0}{2R^2}x. \tag{9.23}$$

A $\cos p\Theta$ configuration with circumferential line current density exists only in theory and would have to be approximated. A constant overall current density j_{av} coil cross-section limited by two horizontal ellipses as shown in Fig. 9.3 will also yield an ideal dipole:

$$B_y = -\frac{\mu_0 j_{av} s c}{b + c}; \quad B_x = 0. \tag{9.24}$$

For the special case of two intersecting circles one obtains

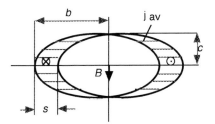

FIG. 9.3. Constant j_{av} horizontal intersecting ellipse dipole magnet configuration.

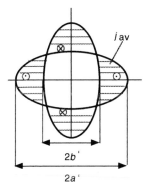

FIG. 9.4. Constant j_{av} intersecting orthogonal ellipse quadropole magnet configuration.

$$B_y = -\frac{\mu_0 j_{av} s}{2}. \tag{9.25}$$

For a quadrupole configuration, delimited by two identical orthogonal ellipses according to Fig. 9.4, one obtains for the field gradient [4]

$$g = \mu_0 j_{av} \frac{a' - b'}{a' + b'}. \tag{9.26}$$

Confronted with the practical aspects of the intersecting ellipse configuration, mechanical design problems must be solved. Figure 9.5 shows an ellipse-like approximation by rectangular coil blocks. Such considerations led to the nowadays predominantly used coil configuration with one or several concentric layers, wound with cables of average current density $j[\text{Am}^{-2}]$ and limited by radii R_1, R_2 and angles Φ_1, Φ_2. R_s is

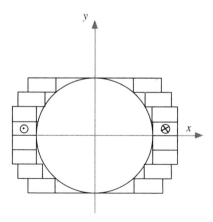

FIG. 9.5. Block approximation of an intersecting ellipse dipole winding configuration.

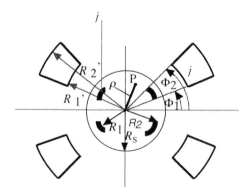

FIG. 9.6. Constant j sector winding with iron screen.

the radius of a concentric iron screen of average relative permeability μ_r; see Fig. 9.6. The vector potential of such a sector winding at the point $P(\rho, \Phi)$, neglecting the iron contribution, is equal to [5]:

$$A(\rho, \Theta) = \frac{\mu_0 j}{\pi} \sum_{n=1}^{\infty} \frac{2p\rho^n \cos n\Theta}{n^2} \sin n\Phi \sum_{R_1}^{R_2} \frac{dr}{r^{n-1}} \qquad (9.27)$$

$$B_\Theta = -\frac{2\mu_0 j}{\pi} \frac{p \, \rho^{n-1}}{n} \frac{\sin n\Phi}{\cos n\Theta} \cos n\Theta \sum_{R_1}^{R_2} \frac{dr}{r^{n-1}} \tag{9.28}$$

$$B_\rho = -\frac{2\mu_0 j}{\pi} \frac{p \, \rho^{n-1}}{n} \frac{\sin n\Phi}{\sin n\Theta} \sin n\Theta \sum_{R_1}^{R_2} \frac{dr}{r^{n-1}}. \tag{9.29}$$

Imaging the coil sector on the iron screen and assuming $\mu_r = \infty$ one obtains:

$$R_1' = \frac{R_s^2}{R_1}; \; R_2' = \frac{R_s^2}{R_2}; \; \Phi' = \Phi \tag{9.30}$$

and for the imaged current density

$$j = j \frac{R_1^2 R_2^2}{R_s^4}. \tag{9.31}$$

Integrating eqns (9.28) and (9.29) one obtains the following expressions for B_ρ and B_Θ within the aperture for $\rho < R_1$, within the coil for $R_1 < \rho < R_2$, and outside the coil for $\rho > R_2$ [5]:

For a **dipole** within the aperture:

$$\begin{Bmatrix} B_\Theta \\ B_\rho \end{Bmatrix} = -\frac{2\mu_0 j}{\pi} \left\{ (R_2 - R_1) (\sin \Phi_2 - \sin \Phi_1) \begin{Bmatrix} \cos \Phi \\ \sin \Phi \end{Bmatrix} \right.$$

$$+ \sum_{n=1}^\infty \frac{\rho^{2n}}{(2n+1)(2n-1)} \left(\frac{1}{R_1^{n-1}} - \frac{1}{R_2^{n-1}} \right) [\sin(2n+1)\Phi_2 \tag{9.32}$$

$$\left. - \sin(2n+1)\Phi_1] \begin{Bmatrix} \cos(2n+1)\Theta \\ \sin(2n+1)\Theta \end{Bmatrix} \right\}$$

and for the iron contribution:

$$\begin{Bmatrix} B_{\Theta Fe} \\ B_{\rho Fe} \end{Bmatrix} = -\frac{2\mu_0 j}{\pi} \frac{\mu_r - 1}{\mu_r + 1} \left\{ \sum_{n=1}^\infty \frac{\rho^{2n-2}}{R_s^{4n-2}(2n+1)(2n-1)} \right.$$

$$\times [R_2^{2n+1} - R_1^{2n+1}][\sin(2n-1)\Phi_2 - \sin(2n-1)\Phi_1] \tag{9.33}$$

$$\left. \times \begin{Bmatrix} \cos(2n-1)\Theta \\ \sin(2n-1)\Theta \end{Bmatrix} \right\}.$$

For the field components within the winding $R_1 < \rho < R_2$:

$$\begin{Bmatrix} B_{\Theta w} \\ B_{\rho w} \end{Bmatrix} = -\frac{2\mu_0 j}{\pi}(R_2 - \rho)(\sin\Phi_2 - \sin\Phi_1)\begin{Bmatrix} \cos\Theta \\ \sin\Theta \end{Bmatrix}$$

$$\mp \sum_{n=1}^{\infty}\left[1 - \left(\frac{R_1}{\rho}\right)^{2n+1}\right]\left[\frac{\rho}{(2n+1)(2n-1)}[\sin(2n-1)\Phi_2 \quad (9.34)\right.$$

$$\left. - \sin(2n-1)\Phi_1\right] \times \begin{Bmatrix} \cos(2n-1)\Theta \\ \sin(2n-1)\Theta \end{Bmatrix}.$$

The iron screen contribution according to eqn (9.33) has to be added to this expression. For the field components outside the winding $\rho > R_2$ one has:

$$\begin{Bmatrix} B_{\Theta e} \\ B_{\rho e} \end{Bmatrix} = \frac{2\mu_0 j}{\pi}\sum_{n=1}^{\infty}\frac{\rho}{(2n-1)(2n+1)}\left[\left(\frac{R_2}{\rho}\right)^{2n+1} - \left(\frac{R_1}{\rho}\right)^{2n+1}\right] \quad (9.35)$$

$$\times [\sin(2n-1)\Phi_2 - \sin(2n-1)\Phi_1]\begin{Bmatrix} \cos(2n-1)\Theta \\ \sin(2n-1)\Theta \end{Bmatrix}$$

where the iron contribution has again to be added. For a **quadrupole** configuration one obtains: for the field components within the aperture $\rho < R_1$

$$\begin{Bmatrix} B_{\Theta} \\ B_{\rho} \end{Bmatrix} = -\frac{2\mu_0 j}{\pi}\left\{\rho\ln\frac{R_2}{R_1}[\sin 2\Phi_2 - \sin 2\Phi_1]\begin{Bmatrix} \cos 2\Theta \\ \sin 2\Theta \end{Bmatrix}\right.$$

$$+ \sum_{n=1}^{\infty}\frac{\rho}{2n(4n+2)}\left[\left(\frac{\rho}{R_1}\right)^{4n} - \left(\frac{\rho}{R_2}\right)^{4n}\right]\sin(4n+2)\Phi_2 \quad (9.36)$$

$$\left. - \sin(4n+2)\Phi_1]\begin{Bmatrix} \cos(4n+2)\Theta \\ \sin(4n+2)\Theta \end{Bmatrix}\right\}.$$

The contribution of the iron screen is:

$$\begin{Bmatrix} B_{\Theta Fe} \\ B_{\rho Fe} \end{Bmatrix} = -\frac{2\mu_0 j}{\pi}\frac{\mu_r - 1}{\mu_r + 1}\left\{\sum_{n=0}^{\infty}\frac{\rho^{4n+1}}{2(n+1)(4n+2)R_s^{4(2n+1)}}\right.$$

$$\times \left[R_2^{4(n+1)} - R_1^{4(n+1)}\right][sin(4n+2)\Phi_2 - sin(4n+2)\Phi_1] \quad (9.37)$$

$$\left. \times \begin{Bmatrix} cos(4n+2)\Theta \\ sin(4n+2)\Theta \end{Bmatrix}\right\}.$$

The field components within the winding for $R_1 < \rho < R_2$ are:

$$\begin{Bmatrix} B_{\Theta w} \\ B_{\rho w} \end{Bmatrix} = -\frac{2\mu_0 j}{\pi} \left\{ \rho \ln \frac{R_2}{R_1} [\sin 2\Phi_2 - \sin 2\Phi_1] \begin{Bmatrix} \cos \Theta \\ \sin \Theta \end{Bmatrix} \right.$$

$$\mp \sum_{n=1}^{\infty} \frac{\rho}{2n(4n-2)} \left[1 - \left(\frac{R_1}{\rho} \right)^4 \right] [\sin(4n-2)\Phi_2$$

$$\left. - \sin(4n-2)\Phi_1] \begin{Bmatrix} \cos(4n-2)\Theta \\ \sin(4n-2)\Theta \end{Bmatrix} \right\}$$

(9.38)

plus the contribution of the iron screen. The field components outside the winding for $\rho > R_2$ are:

$$\begin{Bmatrix} B_{\Theta e} \\ B_{\rho e} \end{Bmatrix} = -\frac{2\mu_0 j}{\pi} \sum_{n=1}^{\infty} \rho \left[\left(\frac{R_2}{\rho} \right)^{4n} - \left(\frac{R_1}{\rho} \right)^{4n} \right] \frac{1}{2n(4n-2)}$$

$$\times [\sin(4n-2)\Phi_2 - \sin(4n-2)\Phi_1] \begin{Bmatrix} \cos(4n-2)\Theta \\ \sin(4n-2)\Theta \end{Bmatrix}$$

(9.39)

plus the iron screen contribution.

Constant current density j has so far been assumed. In practice this assumption will be valid in small rectangular coductors. In keystoned superconducting cables one can in a closer approximation assume a linearily decreasing j between the sector winding radii R_1 and R_2. For that case expressions similar to eqns (9.32–9.39) have been worked out [6].

Any practical application of the above equations for dipole, quadrupole, or higher order multipole fields must be considered in relation to the required field precision. In fact, high field precision is required for superconducting dipole and quadrupole magnets with relative field errors of the order of 10^{-4}. To obtain this degree of precision, higher field multipoles must be eliminated by optimizing the winding cross-section. As an example, by choosing a sector winding with $\Phi_2 = \pi/3$; $\Phi_1 = 0$ the third harmonic term can be suppressed. In a two-layer design the third and fifth harmonics can be suppressed, and so on. This approach led to rather tedious trial and error methods using magnetic field computer codes. Nowadays **inverse** field optimization computer codes are used [7,8], whereby a number of important input parameters are imposed on the program, such as the number of layers, the load lines and dimensions of the superconducting cables, the number of blocks and wedges in each layer, certain geometrical limitations, etc. The inverse computer code will then calculate a minimum error winding configuration, as explained in Section 9.6. The derived analytical expressions will, however, be useful in drafting an initial two-dimentional winding geometry, yielding a first estimate of the magnetic field in the aperture and the coils and will help in the mechanical force calculations, where ultimate field precision is not required. Figure 9.7 shows the conductor repartition in an LHC dipole magnet with a two layer sector winding subdivided into blocks by wedges.

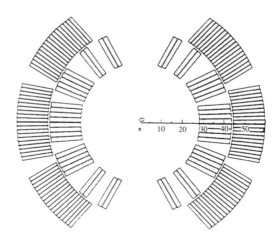

FIG. 9.7. Conductor distribution in a two layer sector winding with wedges of an LHC
dipole magnet.

9.3   End field calculations

The determination of end fields and of their contribution to the main two-dimensional
field in superconducting magnets is different from procedures applied to classical, iron
contour dominated magnets. Here a high degree of uniformity or linearity of the integrated field can be obtained by simple shimming of the pole end faces. One has to bear
in mind that even in 10–15 m long supercoducting dipoles the end field contribution
will amount to a few per cent of the integrated field; at stipulated errors of the field
integral of $\approx 10^{-4}$, the end field multipole errors should not exceed a few per mille.
To this end one can use the **Lambertson criterion** [9] stating that the field integral
distribution depends upon the current integral distribution in the same way as the field
distribution depends on the current distribution in a 2D magnet. The $\sum_{-\infty}^{\infty} j_s \mathrm{d}s$ of a
2D configuration will give rise to end field multipoles. By shifting and subdividing the
winding blocks in the longitudinal direction, one can optimize an end field design and
minimize its higher multipoles. Powerful 3D magnetic field computer codes can also be
used, requiring more data preparation and computation time compared to 2D programs.
An elegant end field calculation and optimization method for sector windings has been
worked out by Iispert *et al.* [10]. The parts contributing to the integrated end field are:
the curved part of the coil end according to Fig. 9.8; the straight part in the end due to
the quasi-constant perimeter winding, see Fig. 9.9; and the additional straight parts due
to end spacers in between adjacent winding blocks, as shown in Fig. 9.10.

The starting point is the expression for the integrated field length in the **straight** part
of length l[m]:

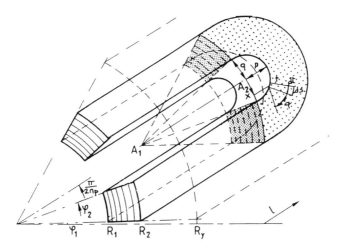

FIG. 9.8. Model for the calculation of the curved end part contribution.

$$\int_0^1 (B_\Theta dl)_{\Theta=0} = \frac{2\rho\mu_0 j}{\pi n} l \Big[(\sin n\Phi_2 - \sin n\Phi_1)$$
$$\times \left(\frac{R_2^{2-n} - R_1^{2-n}}{2-n} + \frac{1}{R_s^{2n}} \frac{R_2^{2+n} - R_1^{2+n}}{2+n} \right) \Big] l = b_{1n} + b_{2n} \quad (9.40)$$

with $2p$ the number of poles and b_{1n}, b_{2n} the coil and iron screen contribution at an assumed $\mu_r = \infty$. Equation (9.40) corresponds to (9.32) and (9.33) and to (9.36) and (9.37) for the field in the aperture of sector winding dipole and quadrupole magnets.

The **curved** part of the coil end, shown in Fig. 9.9, is bent around the central axis $A1$–$A2$ whereby the coil inner face forms an ellipse with the parameters p_e, q_e. The geometry of the curved part at a coil radius R is given by the inner ellipse parameter x as:

$$\varphi(\alpha, x, t) = \frac{\pi}{2p} \frac{x + t \cos \alpha}{R} \quad (9.41)$$

$$j(\alpha) = j \cos \alpha. \quad (9.42)$$

The coil and iron screen contributions of the curved part are:

$$b_{1n} = \frac{2\mu_0 j p}{\pi} \left[F_2\left(e, \frac{n}{p}, \varphi_2\right) - F_1\left(e, \frac{n}{p}, \varphi_1\right) \right] \frac{1}{n^2} \times \frac{R_2^{3-n} - R_1^{3-n}}{3-n} \quad (9.43)$$

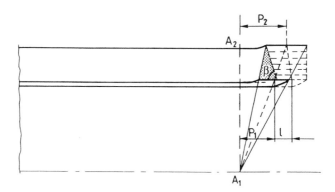

FIG. 9.9. Model for the calculation of the straight part contribution in a constant perimeter winding.

$$b_{2n} = \frac{2\mu_0 j p}{\pi} \left[F_2 \left(e, \frac{n}{p}, \varphi_2 \right) - F_1 \left(e, \frac{n}{p}, \varphi_1 \right) \right] \frac{1}{n^2 R_s^{2n}} \times \frac{R_2^{3+n} - R_1^{3+n}}{3+n} \qquad (9.44)$$

with $e = p_e/q_e$ the ellipse parameter. The numerical values for $F(e, n/p, \Theta)$ are given in the appendix of [10].

The contribution of the **additional straight** part for a constant perimeter winding is due to the difference between the winding length on the top and bottom coil surfaces in the curved end part. This difference Δl is added to the inner coil radius R_1. Since one cannot obtain a constant perimeter configuration throughout a coil block, Δl is calculated at the average radius $R = (R_1 + R_2)/2$. The coil and iron screen contributions are:

$$\begin{Bmatrix} b_{1n} \\ b_{2n} \end{Bmatrix} = \frac{2 p \mu_0 j}{\pi} \Bigg\{ (\sin n\varphi_2 - \sin n\varphi_1) \frac{\pi}{4}$$

$$\times \left[\left(\frac{\pi}{2p} - \varphi_2 \right)(e-1) + \varphi_2 - \varphi_1 \right] \qquad (9.45)$$

$$\times \left\{ \begin{aligned} & \frac{R_2^{3-n}}{2-n} - \frac{R_2^{3-n}}{3-n} - \frac{R_2 R_1^{2-n}}{2-n} + \frac{R_1^{3-n}}{3-n} \\ & \frac{1}{R_s^{2n}} \left(\frac{R_2^{3+n}}{2+n} - \frac{R_2^{3+n}}{3+n} - \frac{R_2 R_1^{2+n}}{2+n} + \frac{R_1^{3+n}}{3+n} \right) \end{aligned} \right\} \Bigg\} .$$

The inclination angle β of the coil end inside face is

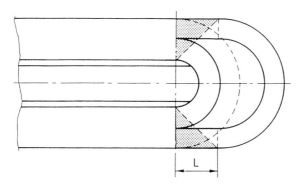

FIG. 9.10. Model for the computation of the end spacer contribution to the straight part.

Table 9.1 *The end field third harmonic content in the 250 Tm^{-1}, 2.5 m long 6 cm bore quadrupole magnet*

	Contribution due to		
Block	Rounded off part	Constant perimeter Δl	Additional straight part
1	-2.2×10^{-5}	-7.2×10^{-5}	3.5×10^{-5}
2	-2.3×10^{-5}	-2.0×10^{-5}	-0.6×10^{-5}
3	1.4×10^{-5}	21.2×10^{-5}	4.2×10^{-5}
4	-0.8×10^{-5}	-0.2×10^{-5}	-0.4×10^{-5}
5	-0.07×10^{-5}	-0.05×10^{-5}	-0.04×10^{-5}
6	0.3×10^{-5}	2.5×10^{-5}	0.6×10^{-5}

$$\operatorname{ctg}\beta = \left(\frac{\pi}{2p} - \varphi_2\right)\left(\frac{\pi}{2}\frac{1+e}{2} - e\right)\frac{\pi}{4}(\varphi_2 - \varphi_1). \qquad (9.46)$$

The contribution of the straight part due to distributed spacers in the coil ends can be computed in a similar way [10].

As an example the end field optimization for a 6 cm bore, 2.5 m long, 250 T m^{-1} field gradient quadrupole magnet is presented. The current density in the 16×2.53 mm^2 keystoned and insulated cable of the two layer winding at maximum field amounts to $j_{9.5T} = 4.56 \times 10^8$ [Am^{-2}]. Figure 9.11 shows the two-dimensional winding cross-section, subdivided into six blocks **1–6**. Figures 9.12a and 9.12b show the optimized coil end configurations for the outer and inner layer, while Table 9.1 summarizes the individual contributions to the third harmonic or twelve-pole in the magnet ends. In terms of the main quadrupole field the third harmonic amounts to 1.08×10^{-4}; at an assumed end field length of 0.25 m and a quadrupole length of 2.5 m, the relative twelve-pole error will amount to 0.44×10^{-4}. The higher harmonic content is considerably smaller.

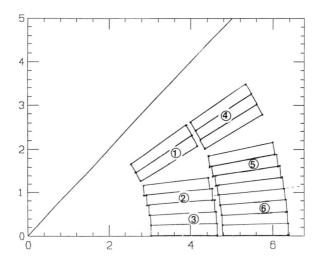

FIG. 9.11. 2D winding cross-section of a 6 cm bore, 2.5 m long supconducting quadropole.

9.4   Field harmonics due to magnetization currents in the superconductor

According to the Bean model for a 'hard' Type II superconductor the conduction of a transport current I_t is always accompanied by the flow in opposite direction of two magnetization currents of critical current density $\pm j_c$, depending on the local B_w. In every superconducting filament the magnetizaton currents thus form elementary dipoles; integrated over the winding cross-section, they will distort the main field pattern. In symmetric multipole magnets odd field harmonics will appear [11,12].

As shown in Fig. 9.13, the elementary dipoles can be expressed by the magnetic moment $M[\mathrm{Am}^{-1}]$. For a rectangular or round filament of width, respectively, diameter d_f, one obtains:

$$\mu_0 M_{\mathrm{slab}} = \frac{1}{4}\mu_0 j_c d_f \tag{9.47}$$

$$\mu_0 M_{\mathrm{circ}} = \frac{2}{3\pi}\mu_0 j_c d_f \tag{9.48}$$

.

As M depends upon the orientation of B_w and on $j_c(B_w)$, the field map of the winding has to be established and the related j_c values found from the characteristic load line of the superconductor. For practical computations the sector coil limited by R_1, R_2, Φ_1, and Φ_2 is then subdivided into n sectors with a surface $\delta F_n = \rho_n \delta \rho_n \delta \varphi_n$ and with α_n

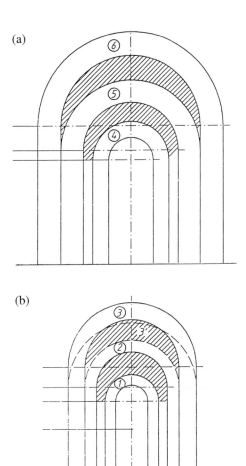

FIG. 9.12. Optimized coil end configuration of the preceeding quadropole;
cross-section at $(R_1 + R_2)/2$.

the orientation of $\vec{B}_\mathrm{w}$, assumed constant in each sector. The harmonics ΔB_{yn} due to M
at point $P_n(\rho_n, \varphi_n)$ at distance ρ on the horizontal axis are:

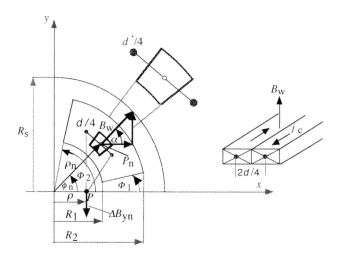

FIG. 9.13. Elementary dipoles due to magnetization currents in sector windings resulting in perturbing field harmonics.

For a **dipole** sector wiring with a concentric iron screen at R_s:

$$
\Delta B_y = \sum_{n=1}^{N} \frac{4\mu_0 M \Delta F_n}{\pi} \sum_{k=0}^{\infty} \left\{ \frac{1}{\rho_n^2} \left(\frac{\rho}{\rho_n} \right)^{2k} \right.
$$

$$
\times \sin[(2k+2)\varphi_n - \alpha] + \frac{\mu_r - 1}{\mu_r + 1} \frac{1}{R_s^2} \left(\frac{\rho\rho_n}{R_s^2} \right)^{2k}
$$

$$
\left. \sin[2k\varphi_n + \alpha] \right\} \left\{ \begin{matrix} 0.85 \\ 1.0 \end{matrix} \right\} .
$$

(9.49)

For a **quadrupole** sector winding one obtains:

$$
\Delta B_y = \sum_{n=1}^{N} \frac{8\mu_0 M \Delta F_n}{\pi} \sum_{k=0}^{\infty} \left\{ \frac{1}{\rho_n^2} \left(\frac{\rho}{\rho_n} \right)^{4k+1} \right.
$$

$$
\times \sin[(4k+3)\varphi_n - \alpha] + \frac{\mu_r - 1}{\mu_r + 1} \frac{1}{R_s^2} \left(\frac{\rho\rho_n}{R_s^2} \right)^{4k+1}
$$

$$
\left. \times \sin[(4k+1)\varphi_n + \alpha \left\{ \begin{matrix} 0.85 \\ 1.0 \end{matrix} \right\} \right\} .
$$

(9.50)

The factor 0.85 stands for round filaments.

The magnetization multipoles will notably be perturbing at low dipole and quadrupole fields where j_c is high. This situation is encountered at low injection energies equivalent to low fields in superconducting accelerators or colliders. To minimize the effect, cables with small filament diameters should be used; in general correcting multipole magnets will have to be provided. Time dependent or dynamic magnetization effects have also been observed and described [13,14]. Although not yet entirely explained, the dynamic effects can be reduced by optimizing the excitation and de-excitation cycle of the superconducting magnets and by using correcting multipole magnets with programmed, time dependent excitation.

9.5   Field errors due to manufacturing and assembly tolerances

Ideal excitation coil and iron screen configurations have been assumed so far. In practice a superconducting magnet consists of many components, manufactured and assembled with a number of geometrical, mechanical, electrical and other tolerances. These deviations will introduce two error groups of a **systematic** and **random** nature which can again be expressed in terms of perturbing field multipoles. Two cases of systematic errors have already been mentioned, the iron screen saturation at high magnetic fields and the magnetization current effect at low inductions. Further systematic errors may be caused by small shifts of one or several sector windings due to punching tool errors, by an excentricity between coils and the iron screen, systematic magnet tilt, and similar effects.

Random effects are caused by fabrication tolerances of the superconductive cable, the insulation thickness, and of other structural elements, by the coolant temperature variation in one magnet or along a magnet string. Systematic errors can be measured and compensated, in contrast to random error. Hence the need for small random deviations or for tight manufacturing and assembly tolerances.

Introducing the errors into magnetic field computer codes the resulting direct and skew field multipoles can be obtained. The inverse process is also important: to determine from the measured field multipoles the most probable geometrical and other tolerances or errors in a specific magnet. An elegant first order perturbation theory for conductor dominated superconducting magnets has been worked out by K. Halbach [15]. To this end, the conjugate complex induction $B^* = B_x - i B_y$ is derived from the complex potential

$$\mu_0 F = \mu_0 \vec{A} + i \mu_0 V \tag{9.51}$$

in the complex $z = x + iy$ plane normalized with respect to $\rho : x = (\xi/\rho); y = (\eta/\rho)$, $d\sigma = (d\xi d\eta)/\rho^2; x, y, z, d\sigma$ are thus dimensionless units. For B^* one obtains:

$$\rho B^* = i\mu_0 \frac{dF}{dz}. \tag{9.52}$$

Expanding $F(z)$ into a Taylor series:

$$F(z) = \sum_{n=0}^{\infty} \mu_0 C_n z^n \tag{9.53}$$

$$B^* = \sum_{n=1}^{\infty} \frac{i\mu_0 C_n}{\rho} z^{n-1} = \sum_{n=1}^{\infty} \mu_0 c_n z^{n-1}. \tag{9.54}$$

For a filament at z_0 carrying a current I the complex potential is:

$$\mu_0 F(z_0) = -\frac{I\mu_0}{2\pi} \ln\left[(z_0 - z) \cdot \left(z_0 - \frac{R_s^2}{z^*}\right)\right]. \tag{9.55}$$

The magnetic induction for a distributed current density is:

$$B^*(z_0) = \frac{i\rho\mu_0}{2\pi} \int j \left(\frac{1}{z - z_0} + \frac{1}{\frac{R_s^2}{z^*} - z_0}\right) d\sigma. \tag{9.56}$$

By a Taylor expansion around z_0 one obtains:

$$a_n = \frac{i\rho\mu_0}{2\pi} \int z^{-n} j \, d\sigma \tag{9.57}$$

$$b_n = \frac{i\rho\mu_0}{2\pi} \int z^{*n} j \frac{d\sigma}{R_s^2} \tag{9.58}$$

$$c_n = a_n + b_n. \tag{9.59}$$

and the normalized iron screen radius R_s. For a symmetric $2p = 2N$-pole structure, invariant for a π/N rotation, one obtains:

$$c_n = c_n \sum_{m=0}^{2N-1} \exp\left[-im\pi\left(1 + \frac{n}{N}\right)\right] \tag{9.60}$$

yielding, for $n = N(2m + 1)$:

$$c_n = 2N c_n; \, b_n = 2N b_n; \, a_n = 2N a_n \tag{9.61}$$

and for $n \neq N(2m + 1)$

$$\underline{c}_n = \underline{a}_n = \underline{b}_n = 0. \tag{9.62}$$

Coming now to some basic errors in a symmetric $2N$-pole magnet, if all sectors of the excitation coil have the same perturbations Δa_n, Δb_n, Δc_n the effect is obtained by introducing them into eqn (9.60) for a_n, b_n, c_n If a particular sector n is **rotated** by a small angle α the first order effect is:

$$\Delta a_n = -in\alpha_n a_n; \ \Delta b_n = -in\alpha b_n \tag{9.63}$$

For a **conductor contour** perturbation at a fixed excitation current I one obtains:

$$\Delta a_n = \frac{i\rho\mu_0}{2\pi} j\Delta \left(\int z^{-n} d\sigma \right) - a_n \frac{d\sigma}{\sigma} \tag{9.64}$$

$$\Delta b_n = \frac{i\rho\mu_0}{2\pi R_s^{2n}} j\Delta \left(\int z^{*n} d\sigma \right) - b_n \frac{d\sigma}{\sigma}. \tag{9.65}$$

For a single conductor or cable **displacement** Δz one has to replace in eqn (9.57) the term z^{-n} by $(z + \Delta z)^{-n}$; the first order expansion yields:

$$\Delta a_n = -n\Delta z \int z^{-(n-1)} d\sigma = -n\Delta z \, a_{n+1} \tag{9.66}$$

$$\Delta b_n = n\Delta z^* \frac{b_{n-1}}{R_s^2}. \tag{9.67}$$

For a displacement by $-\Delta z$ of the **iron screen** with respect to the centre of the excitation coils one obtains:

$$\Delta c_{N(2m+1)-1} = \Delta z b_{N(2m+1)}[N(2m+1) - 1] \tag{9.68}$$

$$\Delta c_{N(2m+1)+1} = \Delta z^* \frac{1}{R_s^2} b_{N(2m+1)}[N(2m+1) + 1]. \tag{9.69}$$

9.6 Computer codes for magnetic field computations

It has already been stated that computer programs for magnetic field computations can be divided into two groups, a larger group of **direct** programs, which will calculate the repartition of the magnetic field, the field harmonics, and the mechanical forces for a given coil and iron screen configuration, and into a smaller group of inverse programs which will, starting from a number of basic input parameters and imposed limiting conditions, calculate optimized winding configurations for dipoles, quadrupoles, and higher multipole magnets exhibiting a minimum or zero field harmonic errors. By combining inverse and direct computer codes solutions and answers to the following problems are obtained: design of economic coil configurations, possibly of minimum superconductor volume, the 2D and 3D repartition of the magnetic induction B and of the integrated field $\int_{-\infty}^{+\infty} B(\rho, \Theta)\mathrm{d}l$, computation of field harmonics, iron saturation and its compensation, magnetization current effects, EM forces, and eddy current effects in the coils and structural parts.

As to the build-up, the architecture of a specific computer program, one has to distinguish between 2D and 3D codes, the type of potential the program is based upon, the program formulation such as the finite difference, finite element, and integral formulation methods, the methods of discretization applied to solve the matrix equations, the rapidity of the program, and the yield of output quantities; furthermore the absolute and relative precision of the calculated magnetic field, the program graphical capabilities, and finally the computer environment and user friendliness. The choice of the vector or scalar potential is less important for 2D-programs; scalar and vector potential based 3D-programs are more complex and imply more unknowns per mesh point and more computer time. Programs for time-dependent fields use both potentials. Modern programs must be able to handle materials of linear and non-linear permeability, isotropic and anisotropic, as well as diamagnetic materials. The methods for solving the matrix equations depend on the number of unknown elements n: for $n < 1000$ direct methods can be used, for $n > 1000$ overrelaxation methods are applied, while conjugate gradient and ICCG (incomplete Choleski conjugate gradient) methods are used in modern finite element programs because of their fast convergence. The discretization for finite element methods should allow us to adapt the meshing network to the configuration of different regions. 2D programs have linear or quadratic triangular or rectangular elements. Most modern programs have automatic meshing routines.

Graphical representation is of considerable importance and help throughout the programing process, notably for data verification. Flux-line plots will help in the verification of the correct selection of boundary conditions. An advanced program must have the capability of extracting and presenting any relevant result in a numerical or/and graphical form.

In what follows a list of some commonly used magnetic field programs and their characteristics is given.

MAGNET 2D program. Finite difference method used; based on scalar potential. Uses different versions for different mesh sizes; harmonic analysis included.
Author: Ch. Iselin, CERN, Geneva, Switzerland.

POISSON, POISCR. 2D program. Based on finite element method, triangular mesh used. Conjugate gradient method applied to matrix equation; program developed for magnetic field, electrostatic, and temperature problems in cartesian and polar coordinates. The program accepts non-linear and anisotropic materials. For stepwise execution, subprograms Automesh, Lattice, Poisson, Triplot and Force are used. Optimization routines for shaping the iron screen to minimize the saturation

harmonics are incorporated.
Authors: K. Halbach, LBL, USA, R. Holsinger and Ch.Iselin, CERN.

FLUX 2D and 3D. Based on finite element method with triangular and rectangular mesh elements in cartesian and polar coordinates. The program has several modules for solving electro- and magnetostatic, electromagnetic, and thermal flow problems in their static, transient, and dynamic state as well as combined electromagnetic and thermal problems. The ICCG matrix solution method is used. The program accepts nonlinear and anisotropic materials. A powerful graphic pre- and postprocessing tool is incorporated. Both programs have CAD interfacing facilities.
Author: CEDRAT S.A., 38240 Meylan, France.

ANSYS. Program primarily developed for mechanical stress analysis. The program was later extended to static and low frequency electromagnetic problems with 2D and 3D capabilities. It offers the possibility of automatic design optimization.
Authors: Swanson Analysis Systems, USA.

OPERA 2D and 3D (formerly GFUN). Program based on the integral method. Discretization used in iron parts; fields created by excitatin coils solved by Biot–Savart's rule. Scalar plus reduced scalar potential applied in the conductor regions. Preconditioned conjugate gradient method for solving the matrix equations used. The program has pre- and postprocessors, with powerful interactive graphic routines for data entry and evaluation of results. Program uses cartesian and polar coordinates.
Authors: B. Trowbridge and J. Simkin; program sold commercially by Vector Fields Ltd, Oxford, Great Britain.

MAFIA (formerly PROFI). Finite difference method applied. Program uses scalar or vector potential. Rectangular or cylindrical coordinate meshing. Matrix equation solved by point and block overrelaxation method. Program nowadays used for solving electromagnetic and high frequency problems. The program had initially been developed for electrical machines.
Authors: Prof. Weiland, Technische Hochschule, 6100 Darmstadt, Germany.

MagNet. 2D and 3D finite element analysis program for solving electrostatic, electromagnetic and high frequency (microwave) problems.
Authors: Infolytica, Montreal, Canada, HBA 1M8.

A useful review and appreciation of the above and some more computer codes is given in the IEEE report [17].

With the emergence of **inverse** programs for the optimization of magnetic fields [7,8], and the expansion of their capabilities to rapidly provide optimized coil configurations, the magnet designer nowadays has access to a powerful and indispensable tool. The structure, capabilities, and the application of inverse computer programs will be explained for the CERN ROXIE 2D code (**R**outine for the **O**ptimization of magnet **X**-sections, **I**nverse problem solving and **E**nd region design). The program has been developed for:

(i) the determination of optimized excitation coil cross-sections for superconducting accelerator magnets of low harmonic content, based on a number of input

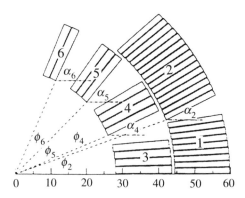

FIG. 9.14. Cross-section of a superconducting LHC dipole model, optimized by the inverse computer code ROXIE.

parameters and on certain imposed limiting conditions.

(ii) the indication of the most probable geometrical errors in the excitation coils and iron screen during fabrication, assembly and cooldown, computed from the measured field multipole content.

(iii) the optimization of a coil end design for a given 2D coil cross-section.

Figure 9.14 shows the coil cross-section of a superconducting dipole magnet, optimized by Roxie. An initial, non-optimized coil geometry is given containing the following parameters: the bare and insulated cable dimensions for the different layers, the number of turns and blocks in each layer, the inclination angles $\alpha_1, \ldots, \alpha_n$, the number of wedges between adjacent blocks and their minimum width, the limiting angles Φ_{lim} for the two innermost blocks 6 and 2, the cable load lines for each layer, the nominal current(s) and others. The iron screen permeability is assumed to be $\mu_r = \infty$. To obtain the relative multipole errors with high accuracy of $b_{(2n+1)k}/b_{1k} \approx 10^{-7} - -10^{-9}$ one proceeds to a fine discretization of the cable cross-section by subdividing it into $N_1 N_2 \approx 100$ current elements of variable j taking also the cable keystoning into account.

The mathematical tool for solving the optimization problems is described in the literature [7]. A penalty function method is used, transforming the initially non linear constrained problem into a set of unconstrained ones. The strongly increasing penalty functions attached to higher multipoles are adjusted at the beginning and kept invariant during subsequent optimization runs. The third option of Roxie, the coil end design is largely based upon the computations given in Section 9.3.

10

MECHANICAL COMPUTATIONS AND DESIGN PRINCIPLES FOR SUPERCONDUCTING MAGNETS

10.1 Introduction

A sound and thorough mechanical design kf high field superconducting magnets is of prime importance for their successful operation. In the early days of magnet development the computational and design effort was mainly concentrated on the analysis and realization of suitable field configurations, while the mechanical problems had been treated in a rather marginal way. The poor test results, failures, and premature quenches experienced with these early magnets indicated the need for a thorough understanding, analysis, and solution of the mechanical problems as well. Concentrated efforts in this direction were notably stimulated by important projects of superconducting high energy accelerators and colliders with several hundred or thousand superconducting dipole, quadrupole, and other magnets: the 1000 GeV Tevatron collider of the NFAL laboratory in the USA, the 800 GeV–30 GeV p-e$^-$ collider Hera at DESY, Germany, and the project of the 7 TeV - 7 TeV collider LHC of CERN. These machines (will) consist of a large number of identical superconducting magnets of required reliable and reproducible optimal performance. A safe mechanical design, preventing early training, repeated quenching and degradation, coil geometries providing the required field quality, an efficient magnet cooling system and quench detection and protection system are the pillars of any successful superconductive magnet design. A safe mechanical design is supposed to meet the following requirements:

(i) it should contain the EM forces and the associated stresses in the coils and the structural parts of the magnet. The excitation coils should always be under compression, during assembly at ambient temperature, during cooldown, and during excitation and operation at nominal induction B_0 and current I_0. The excitation coils should not be submitted to tensile stresses.

(ii) the structural elements of the magnet active part should preferably be assembled and adjusted at ambient temperature; further adjustments at cryogenic temperatures should possibly be avoided.

(iii) During temperature cycling and powering of a superconducting magnet, strain and elasticity limits of coils and structural parts should not be exceeded.

10.2 Forces, stresses, and strains in superconducting magnets

Knowledge of the mechanical force and stress distribution in a superconducting magnet is essential for the design of any coherent and efficient mechanical structure. In this chapter the force and stress analysis will mainly concentrate on coaxial and concentric sector winding configurations for predominantly dipole magnets, which exhibit higher stored energies and EM forces than quadrupoles and higher multipole magnets.

The EM forces in a **dipole** sector winding between the radii R_1, R_2 and subdivided into n blocks between the angles Φ_{2n} and Φ_{1n} can be determined from the interaction of the magnetic field components $B_{\rho w}$ and $B_{\Theta w}$ in the winding and the average current density approximated as:

$$j_s = j_{s0} \cos \Theta. \tag{10.1}$$

To calculate the volume forces F_ρ and F_Θ [N m^{-3}] it is sufficient to consider the basic terms of $B_{\rho w}$ and $B_{\Theta w}$ and neglect the higher multipoles. In accordance with eqns (9.33) and (9.34) one obtains:

$$B_{\rho w} = -\frac{2\mu_0 j}{\pi} \sum_{n=1}^{N} (\sin \Phi_{2n} - \sin \Phi_{1n}) \cdot \left\{ (R_2 - \rho) \right. \tag{10.2}$$
$$\left. +\frac{\rho^3 - R_1^3}{3\rho^2} + \frac{\mu_r - 1}{\mu_r + 1} \frac{R_2^3 - R_1^3}{3R_s^2} \right\} \sin \Theta$$

$$B_{\Theta w} = -\frac{2\mu_0 j}{\pi} \sum_{n=1}^{N} (\sin \Theta_{2n} - \sin \Theta_{1n}) \left\{ (R_2 - \rho) \right. \tag{10.3}$$
$$\left. -\frac{\rho^3 - R_1^3}{3\rho^2} + \frac{\mu_r - 1}{\mu_r + 1} \frac{R_2^3 - R_1^3}{3R_s^2} \right\} \cos \Theta.$$

Denoting the right hand side of eqns (10.2) and (10.3) by $f_\rho(\rho)\sin\Theta$ and $f_\Theta(\rho)\cos\Theta$, the coil forces per unit volume are, in Nm^{-3}:

$$F_\rho = -j_s B_{\Theta w} = \frac{2\mu_0 j^2}{\pi} f_\Theta(\rho)\cos^2\Theta \tag{10.4}$$

$$F_\Theta = j_s B_{\rho w} = -\frac{2\mu_0 j^2}{\pi} f_\rho(\rho)\sin\Theta\cos\Theta. \tag{10.5}$$

The azimuthal and radial stresses p_Θ and p_ρ can now be determined assuming that shear stresses are not transmitted and that the coil external surfaces can slide. The stress balance in a winding element of unit length in the s-direction, according to Fig. 10.1 is:

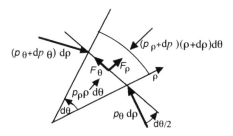

FIG. 10.1. Stress balance in a sector winding element.

$$p_\Theta d\rho - d(\rho p_\rho) + F_\rho \rho d\rho = 0 \qquad (10.6)$$

$$-dp_\Theta + F_\Theta \rho d\rho = 0 \qquad (10.7)$$

Introducing F_ρ and F_Θ into eqns (10.6) and (10.7) one obtains, as worked out by D. Leroy [1]:

$$p_\Theta = -j_s \frac{\rho}{2} \left\{ B_0 \left[\frac{R_2 - \rho}{R_2 - R_1} \frac{\rho^3 - R_1^3}{3\rho^2 (R_2 - R_1)} \right] + B_{Fe} \right\} \cos^2 \Theta \qquad (10.8)$$

$$p_\rho = -j_s \frac{1}{\rho} \left\{ \frac{3}{4} B_{Fe} (\rho^2 - R_1^2) + \frac{3}{4} B_0 R_2 \frac{\rho^2 - R_1^2}{R_2 - R_1} \right.$$
$$\left. - \frac{5}{9} B_0 \frac{\rho^3 - R_1^3}{R_2 - R_1} + \frac{1}{6} B_0 \frac{\rho^3}{R_2 - R_1} ln \left(\frac{\rho}{R_1} \right) \right\} \cos^2 \Theta \qquad (10.9)$$

with B_0, B_{Fe} the bore field and the iron screen contributions:

$$B_0 = -\frac{2\mu_0 j}{\pi} \sum_{n=1}^{N} (\sin\Phi_{2n} - \sin\Phi_{1n})(R_2 - R_1) \qquad (10.10)$$

$$B_{Fe} = -\frac{2\mu_0 j}{\pi} \sum_{n=1}^{N} (\sin\Phi_{2n} - \sin\Phi_{1n}) \frac{\mu_2 - 1}{\mu_r + 1} \frac{R_2^3 - R_1^3}{3R_3^2} \qquad (10.11)$$

The EM forces give rise to bending moments in the coils varying in the azimuthal direction:

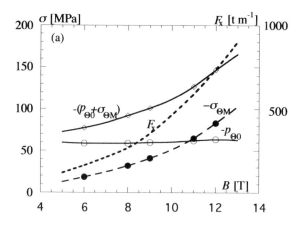

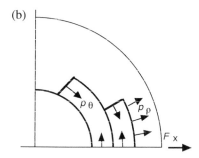

FIG. 10.2. (a) Azimuthal stresses and the EM force F_x on the horizontal midplane in high field dipoles; (b) radial and azimuthal stresses p_r and p_θ and horizontal force F_x in a dipole magnet sector winding.

$$M(\Theta) = -M_{\Theta=0} \cos 2\Theta. \tag{10.12}$$

Along the neutral axis, for $\Theta = \pi/4$, $M = 0$ and for $\pi/4 < \Theta < \pi/2$, $M > 0$. The related azimuthal bending stress σ_Θ is then—see Fig. 10.3 below

$$\sigma_\Theta = \frac{-M \cos 2\Theta}{\dfrac{(\Delta R)^2 \cdot (s = 1)}{6}} \tag{10.13}$$

with ΔR the resulting coil plus retaining structure width and $s = 1$[m] the unit length. In contrast to the always negative compressive stress $-p_\Theta$, σ_Θ becomes a positive tensile

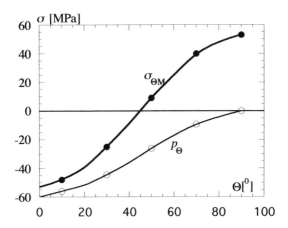

FIG. 10.3. Azimuthal stress distribution in a 10T LHC1m dipole model.

stress for $\Theta > \pi/4$, with a maximum $\sigma_{\Theta m}$ at $\Theta \approx \pi/2$. $\sigma_{\Theta m}$ must be overcompensated by an external azimuthal preload $-p_{\Theta e}$ such that $\mid p_{\Theta e} \mid = \mid \sigma_{\Theta m} \mid + \mid \Delta\sigma \mid$. This condition is a major challenge to the mechanical design, since the total azimuthal compression in the coil on the horizontal axis will then amount to:

$$\sum -p_{\Theta=0} = -p_{\Theta=0} - p_{\Theta e}. \tag{10.14}$$

The cable insulation should withstand this preload; certain superconductors such as Nb_3Sn may experience a j_c reduction due to this preload—see Chapter 5. Close to $\Theta \approx \pi/2$ at the so-called 'poles' the remaining preload is reduced to

$$\sum -p_{\Theta=\frac{\pi}{2}} \approx -\Delta\sigma. \tag{10.15}$$

Figure 10.2 shows the $-\sigma_{\Theta=0}$ and $-\sigma_{\Theta=0} - p_{\Theta=0}$ stresses calculated on the horizontal midplane [2] depending upon the bore field B_0. The almost constant azimuthal compression $-p_{\Theta=0}$ between 5 T and 13 T can be explained by the quasi-constant product of $-p_{\Theta=0} \propto k j_c B_w$. Another explanation is the behaviour of the pinning force $f_p[Nm^{-3}]$ in NbTi and Nb_3Sn, shown in Fig. 2.10. Between 5 T and 13 T, f_p decreases almost in the same way as the superconductor cross-section or unit length volume increases with the field, such that the transverse pressure $-p_{\Theta=0}$ in the winding, due to the vertical EM force $-F_y$, remains quasi-constant. Also shown is the horizontal EM force $F_x = f(B_0)$. Figure 10.3 shows the curves $\sigma_\Theta = f(\Theta)$ and $\sum -p_{\Theta=0} = f(\Theta)$ computed for the 9T LHC dipole of CERN according to Fig. 10. 4a.

So far the EM forces and stresses acting in the transverse plane have been considered. The coil ends are also subjected to longitudinal forces and stresses F_s, σ_s, which

(a)

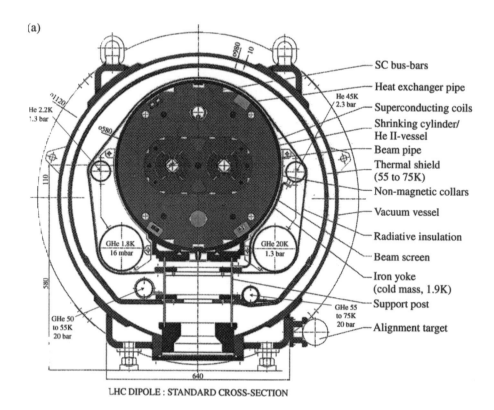

SC bus-bars

Heat exchanger pipe

Superconducting coils

Shrinking cylinder/
He II-vessel

Beam pipe

Thermal shield
(55 to 75K)

Non-magnetic collars

Vacuum vessel

Radiative insulation

Beam screen

Iron yoke
(cold mass, 1.9K)

Support post

Alignment target

LHC DIPOLE : STANDARD CROSS-SECTION

FIG. 10.4A. Cross-section of the 8.4 T bore field, twin aperture LHC dipole (courtesy: CERN, Geneva, Switzerland).

can be determined by assuming the coil ends to be straightened half circles as shown in Fig. 10.5. The bore field at the magnet end B_{0e} is supposed to be known. Depending upon the longitudinal extension of the iron screen one may have $B_{0e} \leq B_0$. The field variation in the axial direction can be approximated by a linear decrease from B_{0e} to zero at the outer coil end. Introducing, in accordance with Fig. 10.5

$$r_1 = \left(\frac{\pi}{2} - \Phi_2\right)\frac{R_1 + R_2}{2}; \quad r_2 = \left(\frac{\pi}{2} - \Phi_1\right)\frac{R_1 + R_2}{2} \qquad (10.16)$$

with R_1, R_2, Φ_1, Φ_2 the 2D parameters of a sector winding, one obtains for the end force F_s[N], acting on a pair (upper and lower) of dipole sector windings [3]:

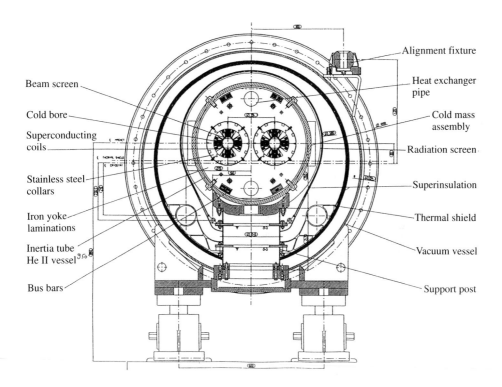

FIG. 10.4B. Cross-section of the 223 T m−1 field gradient, 3.1 m twin aperture main LHC quadrupole.

$$F_s = 4jh \left\{ B_{0e} \frac{r_2^2 - r_1^2}{2} - \Delta B_s \left[\frac{r_2^3 - r_1^3}{3(r_2 - r_1)} - \frac{r_1}{2}(r_1 + r_2) \right] \right\} \qquad (10.17)$$

$$\sigma_s = j \left\{ B_{0e} \frac{r_2 + r_1}{2} - \Delta B_s \left[\frac{r_2^3 - r_1^3}{3(r_2 - r_1)^2} - \frac{r_1}{2} \frac{r_2 + r_1}{r_2 - r_1} \right] \right\} \qquad (10.18)$$

with h the coil height and ΔB_s the field decrease across the coil. A simpler and more elegant method to compute the longitudinal forces F_s consists in the derivation of the magnet stored energy E_m with respect to the longitudinal coordinate s:

$$F_s = \frac{\delta E_m}{\delta s} = \frac{\delta}{\delta s} \left(\frac{LI^2}{2} \right)_{s=1m} \qquad (10.19)$$

$$= \frac{\delta}{\delta s} \left(\sum_{n=1}^{N} \int \frac{B_n^2 (\rho, \Theta)}{2\mu_0} \rho \, d\rho \, d\Theta \right)_{s=1m}. \qquad (10.20)$$

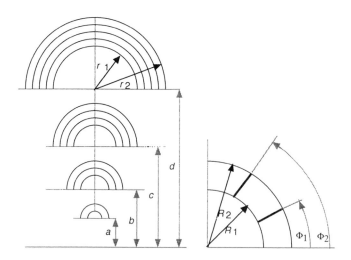

FIG. 10.5. Coil end presentation for end force computation.

In the above equation the magnet energy per unit length $(LI^2/2)_{s=1m}$ or F_s is automatically obtained from the computer code; the stored energy in the magnet can also be obtained from the field map in the $n = N$ regions, comprising the bore field and the fields within and outside the winding layers. To this end eqns (9.32)–(9.39) can be used. In a similar way the radial and circumferential EM forces F_ρ, F_Θ can be obtained from:

$$F_\rho = \frac{\delta E_m}{\delta \rho}, \quad F_\Theta = \frac{1}{\rho} \frac{\delta E_m}{\delta \Theta}. \tag{10.21}$$

10.3   Mechanical design principles

The hybrid, cold iron structure is nowadays predominantly adopted for superconducting high field magnets. The main elements of this structure are shown in Fig. 10.6: the sector coils, wound with insulated cable (1), and placed into laminated collars (2), surrounded by a (split) yoke of magnetic steel (3) and an outer retaining element, an Al or stainless steel cylinder (4). The structure is called hybrid because of the collars, inserted between the coils and the iron. Two arguments speak in favour of such a solution: the iron screen should be placed at a certain distance from the coils in order to reduce saturation and the related field multipoles. Second, the collars are used for preloading the coils at ambient temperature and possibly also during cool-down.

Aiming at achieving an optimal mechanical structure the magnet designer can use a number of constructional elements whereby he can judiciously use their relative contraction coefficients during cool-down $(\Delta l/l)_{293K-4.2K}$ and the elasticity moduli $E[\text{daN} \, \text{mm}^{-2}]$, given in Table 10.1 and shown in Fig. 10.9 below.

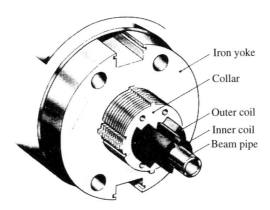

FIG. 10.6. Main cold mass components of a hybrid structure superconducting high field magnet.

Table 10.1 *Thermal contraction and elastic moduli of some basic materials used in superconducting magnets*

Material	Thermal contraction $-\left(\frac{\Delta l}{l}\right)_{293\text{K}-4.2\text{K}}$	Elasticity modulus $E[\text{daN mm}^{-2}]$ at 293 K	at 4.2 K
Insulated, preimpregnated NbTi cable	$3.3 \times 10^{-3} - 3.5 \times 10^{-3}$	$1.5 - 2.5 \times 10^3$	$3 - 4 \times 10^3$
Insulated and impregnated Nb$_3$Sn cable	$2.7 \times 10^{-3} - 2.9 \times 10^{-3}$	$1.5 - 3 \times 10^3$	$2.5 - 4 \times 10^3$
Stainless steel	3×10^{-3}	2.05×10^4	2.15×10^4
Magnetic steel	2.1×10^{-3}	2.05×10^4	2.15×10^4
Copper	$3.1 \times 10^{-3} - 3.3 \times 10^{-3}$	1.3×10^4	1.5×10^4
Aluminium	4.15×10^{-3}	7.5×10^3	8.5×10^3
Bronze	3.8×10^{-3}	1.2×10^4	1.4×10^4
Insulation	$\sim 5 \times 10^{-3}$	1.0×10^3	1.5×10^3

It has already been mentioned that one of the main requirements to be met in super-conducting high field magnets is that of an adequate preloading of the excitation coils. It shall now be discussed how the required preloading can be obtained. Concentric sector windings can be preloaded by mechanical means such as collars, preheated and shrink-fitted or pressure welded cylinders. During cooldown the preload can even be amplified in the case of a positive differential thermal contraction between coils and the clamp-

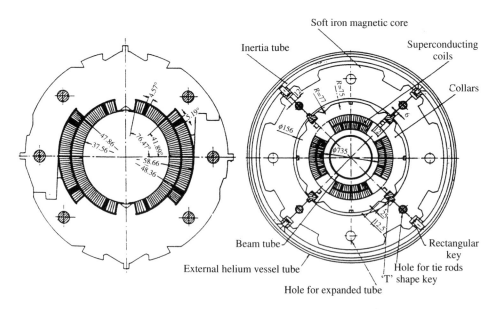

FIG. 10.7. Cross-sections of the main dipole and quadrupole magnets of the HERA collider. (Courtesy: DESY, Hamburg, Germany)

ing system. More sophisticated mechanical structures with airgaps between the mating faces of a usually vertically split iron yoke have also been developed.

For a first estimation, simple shrink-fitting formulae for two concentric rings will be useful. Approximating the excitation winding by a ring between the radii r_1 and r_2 of elasticity E_{12} and the outer collar by a ring between r_2 and r_3 of E_{23}, the reduction δ of the outer ring radius r_2, required to obtain a radial preload $-p_{r2}$ is equal to [4, 5]:

$$\delta = -p_{r2}r_2 \left(\frac{1}{E_{23}} \frac{r_3^2 + r_2^2}{r_3^2 - r_2^2} + \frac{1}{E_{12}} \frac{r_2^2 + r_1^2}{r_2^2 - r_1^2} \right). \tag{10.22}$$

In practice $-p_r$ is obtained by heating and shrink-fitting of the outer ring of $r_2 - \delta$ or by adequately compressing the assembly in a strong press. $-p_{r2}$ is related to the maximum azimuthal preload by:

$$-p_{r2} = -\frac{1}{2} p_{\Theta m} \frac{r_2^2 - r_1^2}{r_2^2}. \tag{10.23}$$

For a coil radius ρ one obtains:

$$-p_\Theta = -p_{r2} \frac{r_2^2}{r_2^2 - r_1^2} \left(1 + \frac{r_1^2}{\rho^2}\right) \tag{10.24}$$

$$-p_r = -p_{r2} \frac{r_2^2}{r_2^2 - r_1^2} \left(1 - \frac{r_1^2}{\rho^2}\right). \tag{10.25}$$

During cool-down and in the case of a positive differential thermal contraction $\delta\,(\Delta l/l) = \delta[(\Delta l/l)_{\text{cyl.}} - (\Delta l/l)_{\text{coil}}] > 0$ such as in case of an Al cylinder, one obtains:

$$-\Delta p_{r2} = \frac{\delta \left(\frac{\Delta l}{l}\right) E \left(1 - \frac{r_2^2}{r_3^2}\right)}{2(1-\nu)} \tag{10.26}$$

with ν the Poisson coefficient and E the average elasticity modulus.

In hybrid magnet structures, according to Fig. 10.6, a negative differential thermal contraction will exist between the collar plus coil part and the surrounding yoke of magnetic steel. In that case the coil plus collar part must be preloaded at ambient temperature; additional preloading may be obtained in the case of $\delta > 0$, such as for Al collars. Superconducting magnets of this type with a closed iron yoke at ambient temperature have been built and successfully operated, such as the HERA collider dipoles of DESY [6, 7], shown in Fig. 10.7, and the prototype dipoles for the abandoned SSC collider project [8], shown in Fig. 10.8.

The vertically split iron yoke design with two gaps (in case the of a dipole) between the mating faces allows one to compensate a negative $\delta < 0$. The upper or lower gap g between the iron halves can be chosen as:

$$Ng \lessgtr r\pi\delta \tag{10.27}$$

with $N = 1, 2, \ldots$, for a dipole, quadrupole, $\ldots$, and r the radius of the collar–yoke mating face. In that case the preloading of the outer cylinder will during cool-down influence the collar plus coil part via the vertically split iron yoke then acting as a force transmitter. For $Ng \geq r\pi\delta$ the gap will not close after cool-down and the EM force F_x will be partly absorbed by the collars and mainly by the outer cylinder. For $Ng \leq r\pi\delta$ the iron faces will close at a temperature $T > T_0$. The remaining differential contraction between T and T_0—see Fig. 10.9—will build up a compressive stress and force $-F_m$ on the vertical faces of the iron yoke; whenever possible one should aim at $|-F_m| \geq |F_x|$. The design can be further refined by inserting in the horizontal slots of the two iron hahves two precisely machined Al bars as shown in Fig. 10.13 below [9, 10]. When the

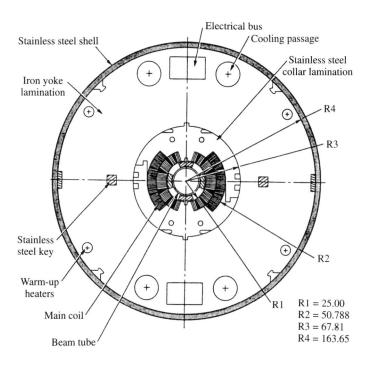

FIG. 10.8. Cross-section of the SSC main dipole with a horizontally split yoke.

gaps are open, the bars are horizontally loaded; during cool-down the bars shrink in a well defined way and control the closing of the gaps.

So far mechanical design aspects of single bore magnets have been treated. Similar considerations also apply to the more sophisticated design of twin aperture LHC dipole and quadrupole magnets, shown in Figs 10.4a and 10.4b. Two variants can be distinguished: the first has two separate coil plus collar parts, placed into a common split iron yoke—see Fig.10.4b; in the second variant the two sets of coils have common collars, as shown in Fig. 10.4a.

So far ideal, friction-free mechanical configurations have been assumed and geometrical tolerances neglected. In order to determine their influence it is advisable to build short and full length mechanical models of nominal cross-section, equipped with adequate measuring devices and sensors like strain gauges. Extensive mechanical measurements at room temperature, during cool-down, and magnet excitation can then be performed and realistic data obtained.

During the design phase, testing or modification of mechanical structures powerful computer codes like ANSYS [11] and CASTEM [12] are applied. Both codes are nowadays extensively used by the comunity of superconducting magnet designers. The pro-

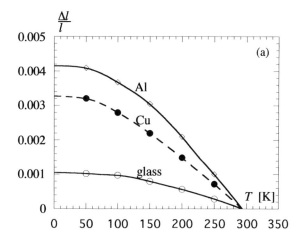

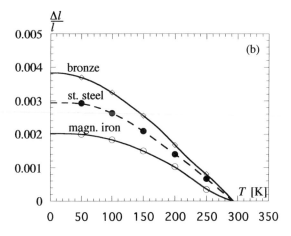

FIG. 10.9. Differential thermal contraction of some important materials used in super-conducting magnets.

grams will solve mechanical, electromagnetic, thermal, or combined problems, related to superconducting magnets. Their versatility has been steadily improved over the last two decades. Material parameters according to Table 10.1 are either permanently stored or can be introduced into the programs. ANSYS can handle contact elements without and with friction, gliding surfaces and materials with thermoelastic, non-linear elastic, and orthotropic properties (like different coil elastic moduli E_Θ and E_r). Electromagnetic, thermal, and mechanical parameters are allocated to each node and elementary mesh surface. A complete mesh may contain several thousand elements, as shown in

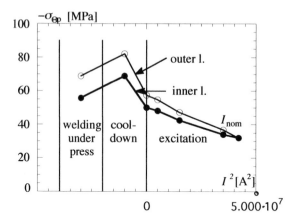

FIG. 10.10. Successive reduction of coil preload at the 'poles' of the SSC dipole.

Fig. 10.16 below. Matrix equations are solved for the three phases of a superconducting magnet, at ambient temperature, cool-down, and during excitation. As a result ANSYS will provide stresses, displacements, and other information in every part of the magnet cross-section. Using automatic mesh generators it should be possible to modify a mechanical concept and to optimize it within a reasonable time.

Examples of some recent NbTi and Nb₃Sn wound high field dipoles which have been successfully built and tested will now be given. As to the mechanical structure, one can distinguish the following arrangements:

The closed iron design; total preload applied at ambient temperature

The coils plus collars, iron yoke, and the outer retaining cylinder are assembled under a press; the required coil preload $-p_{\Theta m}-\varepsilon$ is obtained by welding or bolting the cylinder. ε is the loss of preload when the magnet cold mass is taken out of the press. During cool-down and excitation $-p_{\Theta m}$ is further reduced to the remaining coil prestress $-\Delta\sigma$ at $\Theta = 90°$. This structure had been chosen for the 6.6 T main dipoles of the SSC collider and confirmed by testing a series of prototype magnets. The successive reduction of the initial coil preloading is shown in Fig. 10.10. SSC prototype dipoles had been made with horizontally and vertically split iron yokes. In the first case the stiffness of the iron halves contain the mainly horizontal EM force F_x. In the case of a vertical split the stiff closed yoke will during cool-down prevent the contraction of the cylinder influencing the coil plus collar part in a noticeable way. However a horizontal compression $-F_m = \int -p_m df$ will appear at the iron mating faces to counterbalance F_x. A drawback of the closed iron yoke concept is the required high preload at room temperature where, in accordance with Table 10.1, the elastic moduli of several structural materials are lower than at 4.2 K. The horizontally split yoke concept exhibits a further drawback as the

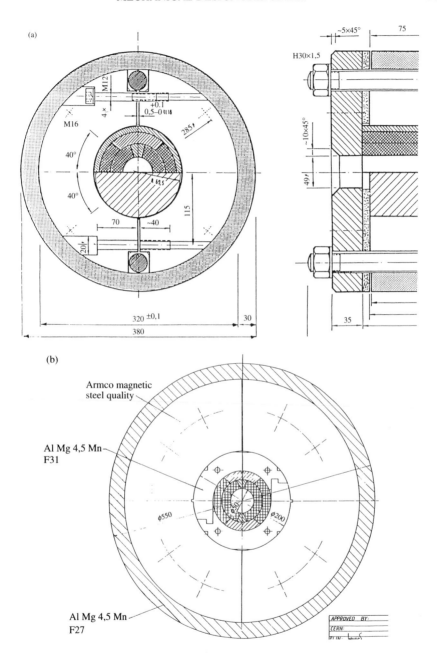

FIG. 10.11. Cross-sections of the 1 m long, Nb_3Sn 10.02 T mirror dipole and of the 9.45 T dipole, made by the CERN–ELIN collaboration. (Courtesy: CERN, Geneva, Switzerland and ELIN, Weiz, Austria

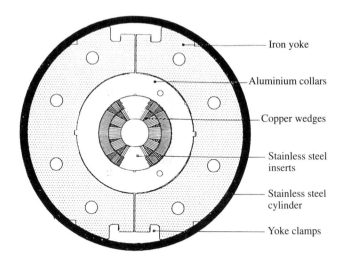

FIG. 10.12. Cross-section of the 1 m, 11.1 T Nb$_3$Sn dipole, made by the Twente University. (Courtesy: Twente University, Netherlands.)

magnetic flux lines then cross the mating faces; the horizontal force F_x may cause a left-right field asymmetry ΔB, resulting in even field harmonics.

Coil preloading is obtained by the collaring system

In superconducting dipoles and quadrupoles with bore fields below $B_0 < 5\text{T}$ the required preload can be obtained at ambient temperature by assembling and keying the collars plus coil part under a press. Using collars of Al alloy the coil compression can be increased during cool-down. The outer retaining ring and the iron yoke do not contribute to the coil preloading. This design has been adopted for the superconducting dipoles (and quadrupoles) of the HERA collider [6, 7]; see Fig.10.7.

The vertically split yoke design with open gaps at room temperature

The principle of this design has already been discussed. The design has been adopted for a number of recent high field dipole magnets: among those wound with NbTi cable, the LHC twin aperture dipoles, shown in Fig. 10.4a, and the D-19 10 T dipole of the LBL [9] are mentioned; among Nb3Sn wound dipoles the first 9.5 T dipole of the CERN–ELIN (Austria) collaboration, shown in Fig. 10.11, [13, 17], the recently successfully tested 11.1 T dipole of the Twente University, shown in Fig. 10.12 [14] and the D-20 13 T dipole of LBL shown in Fig. 10.13 [10] are mentioned.

Let us, as an example for the mechanical concept of dual aperture magnets, discuss the main 15 m long twin aperture dipole for the LHC collider of CERN [15], shown in

Fig. 10.4a. The magnet cold mass consists of the left and right hand two layer excition coils, wound with 15.1 mm wide, insulated NbTi cables around a 5.6 cm aperture and placed into common collars. The coil plus collar package is surrounded by the vertically split yoke and a 1 cm stainless steel cylinder, welded at room temperature under a strong press. The vertical gap is closed at the outer radius of the yoke and open at the inner radius. During cool-down the outer cylinder will contract and the two yoke halves will pivot around the outer contact point before closing. Pressure will thus be applied to the circular collar parts and to the coils. The azimuthal stress balance is:

Total preload during welding of outer cylinder under press	$-p_{\Theta m} - \varepsilon$ =	-120 MPa
Loss of preload due to press release	ε =	$+30$ MPa
Remaining coil preload at 293 K	$-p_{\Theta 293 K}$ =	-90 MPa
Loss of preload during cool-down	$\Delta p_{\Theta cd}$ =	$+25$ MPa
Remaining preload at $T_0 = 1.8$ K	$-p$ =	-65 MPa
Increase of azimuthal preload in the horizontal plane due to EM forces	$-p_{\Theta 0} - \sigma_{M0}$ =	-80 MPa
Total azimuthal coil precompression in the horizontal plane	$\sum -p_{\Theta 0}$ =	-145 MPa
Remaining azimuthal preload at the 'poles' at nominal excitation	$\Delta \sigma$ =	-25 MPa

The merit of the vertical gap design in this example consists in limiting the preload loss during cooldown to only 25 MPa.

The vertically split yoke design has also been adopted for the 1 m long, 13 T prototype dipole of the LBL. As shown in Figs 10.13, the dipole has a four layer graded excitation winding made of rectangular Nb_3Sn cables, wound around the 5 cm bore. The load lines for the two inner and two outer layers are shown in Fig. 10.14. The winding is graded, which means adapted to the respective field maxima. The slim stainless steel collars have a width of only 9 mm. The iron yoke is thus placed close to the coils in order to increase the yoke contribution to the bore field at the price of higher field harmonics due to iron saturation. At ambient temperature the thin collars provide a small preload of -15 MPa. The main preload of -120 MPa is obtained by winding around the two outer stainless steel half shell steel tapes under high tension. At room temperature the tapered gap between the iron halves is adjusted to $\delta = 0.76/0.56$ mm. Figure 10.15 shows the azimuthal coil preload as calculated by ANSYS at 293 K, during cooldown and at nominal field on the horizontal midplane and at the 'poles'. The curves show that

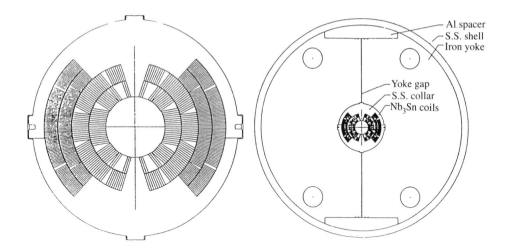

FIG. 10.13. Winding and magnet cross-section of the four layer, 13 T Nb_3Sn dipole of
    LBL (courtesy: LBL, Berkeley, USA).

the weak collars do not provide additional prestressing of the coils during cool-down.
However, the controlled closing of the gaps, monitored by the two horizontal Al bars,
limits the loss of preload to only $+10$ MPa.

Important progress has also been made in the analysis of the end forces, their containment, and in the shaping of the coil ends. Efforts in this direction have, among
others, been stimulated by investigations on the quench origin in a number of high field
superconducting accelerator magnets. It has been found that most quenches occurred
in the coil ends and in the region where the coil straight parts merge into the ends. 3D
computer codes like CASTEM [12] and ROXIE [16] have been developed for shaping
the coil ends according to a quasi-constant perimeter layout—see Fig. 10.17—close to
configurations of 'minimum mechanical energy'. Both codes will also provide data for
precise end piece machining by numerically controlled tooling; Fig. 10.18 shows a set
of such end pieces for the LHC dipole magnet.

The coil end forces can be contained by thick endplates, welded to the outer retaining cylinder or fixed by strong longitudinal bolts passing through holes in the iron yoke,
as shown in Fig. 10.12. The axial preload can be chosen to be smaller or equal to the
total axial force at one magnet end $\sum F_s$. A fraction of $\sum F_s$ is transmitted by friction
from the coils to the outer cylinder which will in return exercise an additional radial
compression on the magnet active part. Magnets with zero axial preload have also been
built and successfully operated.

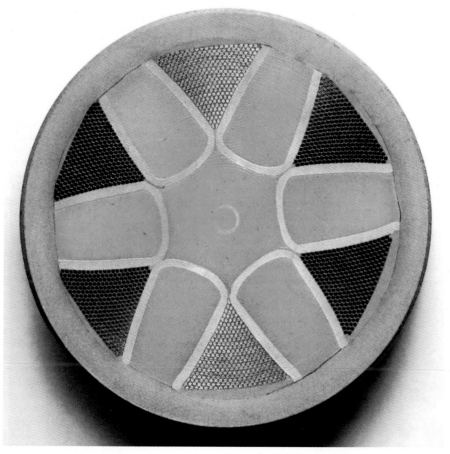

PLATE 1. Example of a mixed matrix NbTi-Cu-CuNi strand of 1.25 mm Φ
(courtesy: Vacuumschmelze GmbH, Germany).

PLATE 2. Testing of the cold diodes for the LHC magnets (courtesy: CERN, Geneva, Switzerland).

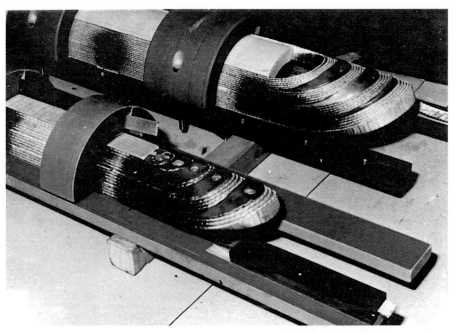

PLATE 3. Picture of the wound inner and outer layers of the 9.45 T ELIN—CERN
Nb_3Sn dipole after reaction (courtesy: ELIN, Weiz, Austria).

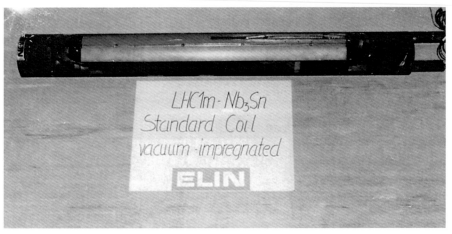

PLATE 4. Picture of the above coils after epoxy resin impregnation in a common
mould (courtesy: ELIN, Weiz, Austria).

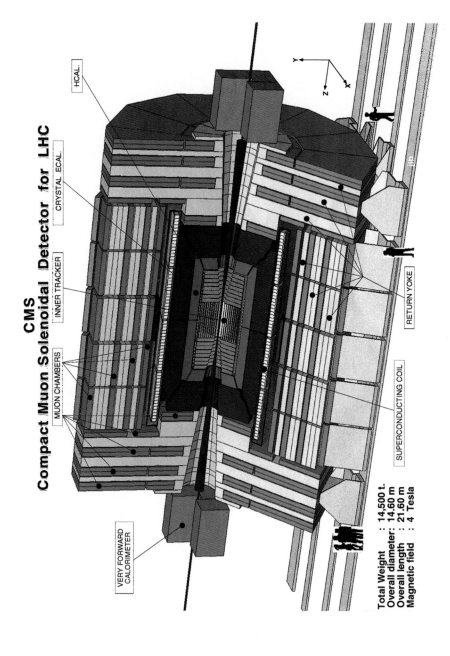

PLATE 5. Main elements of a multipurpose particle detector for colliding beams; example of the CMS detector for the LHC (courtesy: CERN, Geneva, Switzerland).

Sept. 13 90

2-11A of 12

PLATE 6. The Aleph detector solenoid (courtesy: DAPHNIA/STCM, CEA/Saclay, France).

PLATE 7. The fully equipped and installed detector Aleph (courtesy: DAPH-NIA/STCM, CEA/Saclay, France).

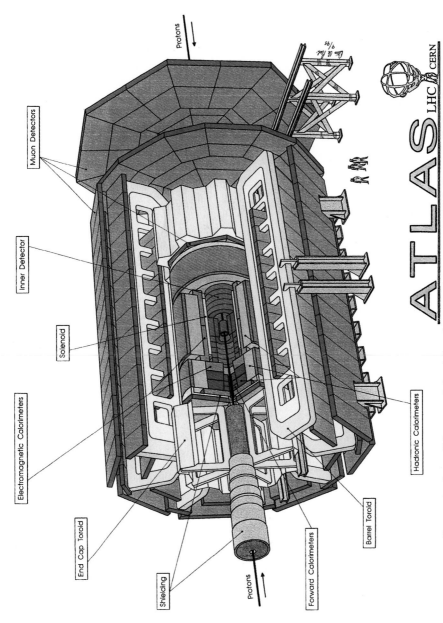

Muon Detectors

Inner Detector

Solenoid

Electromagnetic Calorimeters

End Cap Toroid

Shielding

Protons

Protons

Forward Calorimeters

Barrel Toroid

Hadronic Calorimeters

PLATE 8. Overall layout of the Atlas detector (courtesy: CERN, Geneva, Switzerland).

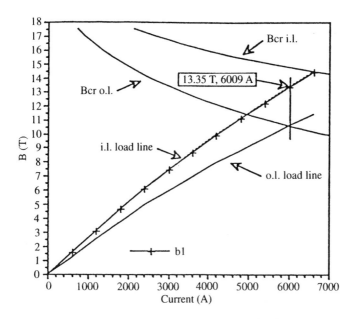

FIG. 10.14. Load lines for the two inner and the two outer layers of the 13 T Nb$_3$Sn LBL dipole.

10.4   Remarks on the final magnet design and the manufacturing aspects

The final layout of a high field superconducting magnet is a complex enterprise which consists in coordinating and streamlining often contradictory requirements and information int a coherent and optimized design. The data refer to the superconductor, insulations, structural materials, temperature and EMF effects, quench behaviour and protection, the mechanical concept, and the cooling system. None of these items should be neglected or introduced at a later design phase. The magnet designer should exhibit solid knowledge in mechanical and electrical engineering, thermodynamics, applied superconductivity and cryogenics. When large series of magnets are involved, economic criteria will also have to be considered. Several items have already been treated in previous chapters. Others are just mentioned without going into more detail, such as punching, machining, stacking or bolting, and keying of mechanical parts, vacuum-tight welding and sealing, to give a few examples.

The relative merits of the two technical superconductors were discussed in Chapter 5. Safety margins should be introduced when defining the nominal working point $P_0(I_0, B_0)$. For NbTi cables a 5–10% degradation due to the cabling process is usually assumed. In addition a 15–20% margin is assumed for P_0 with respect to the intersecting point of the magnet load line and the short sample curve $j_c = f(B)$, as shown in Fig. 10.14. For NbTi magnets operated at 1.8–1.9 K a short sample curve at a higher

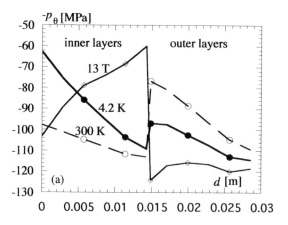

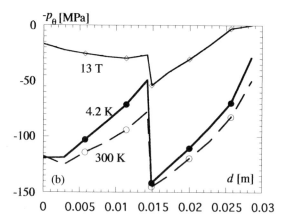

FIG. 10.15. (a) Transverse pressure distribution on the horizontal midplane of the 13.5
T dipole; (b) transverse pressure distribution at the 'poles' of the 13.5 T dipole.

temperature such as $T_{jc} = 2.9 - 3.9 K$ is assumed. If one had chosen $T_{jc} \sim T_\lambda$ the margin with respect to the λ point would amount to only 0.2–0.3 K. Since $j_{c\,2.9K} < j_{c\,1.9K}$
an additional safety factor is obtained. It should, however, be said, that T_λ does not limit
the behaviour of the magnet!

Little degradation has been observed in Nb$_3$Sn cables with small keystoning angles
compared to individual strand measurements. However, any measured critical current
reduction due to transverse pressure $\Delta j_c = f(-p_\Theta)$ must be taken into account—see
Chapter 5. To this end adequate high field and high current test facilities for NbTi and
Nb$_3$Sn cables are required. In graded coils short sample curves, load lines, and working

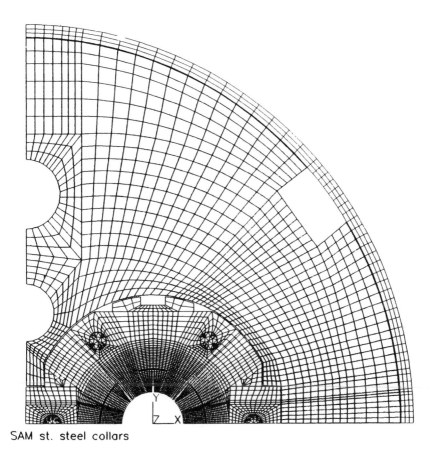

SAM st. steel collars

FIG. 10.16. Example of an ANSYS program mesh for mechanical computations of the
LHC twin aperture dipole magnet.

points must be determined for every layer. It is today assumed that a well designed and
manufactured magnet should reach nominal performance within a few quenches (3–5)
and attain short sample performance after (hopefully) a few more quenches.

The choice of performant insulating materials is rather limited. Let us briefly recall
the expected requirements: high dielectric strength, good mechanical resistance under
compression and against abrasivity, compatibility with epoxy resin impregation, and
reasonable shear stress resistance. A typical two-component **NbTi cable** insulation chosen for the LHC magnets is shown in Fig. 6.33. It consists of two 50% overlapped, 25
μm polyimide (kapton) layers allowing the cable to slightly slide in the longitudinal
direction, and of an outer epoxy-preimpregnated glass-fibre tape wound with a spacing
of $\approx$ 2 mm. The polyimide provides the dielectric strength, the glass-fibre tape the me-

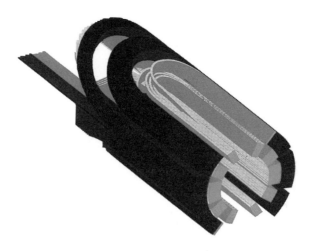

FIG. 10.17. Coil end configuration as calculated for the LHC twin aperture dipole by the ROXIE code.

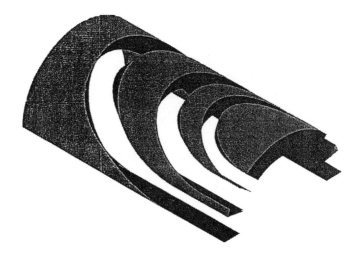

FIG. 10.18. End spacers for the dipole of Fig. 10.17 computed by ROXIE. (Courtesy: CERN, Geneva, Switzerland)

chanical protection. By curing and polymerizing the preimpregnated tape in a precision mould one obtains rather compact coils with channels for helium penetration into the coils. One speaks of partly permeable coils for enhanced cooling and heat evacuation.

The compacted coil is then mass or ground insulated with preformed polyimide

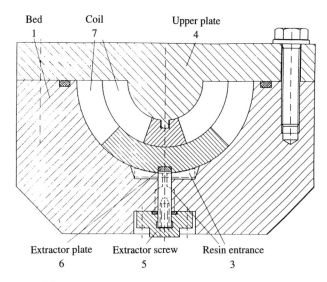

FIG. 10.19. Mould for epoxy resin vacuum impregnation of wound and reacted Nb$_3$Sn coils.

layers, glass web, or layers of glass-epoxy. To reduce friction between coils and adjacent structural elements thin stainless steel or bronze sheets can be inserted.

Coils wound with **Nb$_3$Sn cables** require a different insulation, consistent with the 'Wind and React Technology' developed by A. Asner [17,18]. Due to the small inner radii of the excitation coils of the order of 4–5 mm the winding of coils with brittle, prereacted Nb$_3$Sn cables and a 'classical' insulation cannot be considered. In the 'Wind and React' process the coils are wound with non-reacted, adequately insulated cables which are then reacted and vacuum impregnated with epoxy resin. The insulation must withstand the high reaction temperatures of 700–800°C and still exhibit the required dielectric and mechanical properties. Today two groups of such high temperature resistant insulations are known: high temperature resistant R- or S-glass tapes, sleeves or webs, containing a small amount of organic glue, to make the insulation less brittle. After winding the glue can be washed out or burnt and evacuated at temperatures of 80–100°C. The second group is based on glass-mica tapes and mats with an organic and inorganic glue or binder. The first can be burnt out and evacuated under vacuum at ≈ 600°C during an operation, incorporated into the Nb$_3$Sn reaction process [18]. Both groups of insulations are compatible with epoxy resin impregnation in precision moulds, shown in Fig. 10.19.

For the mass insulation of reacted Nb$_3$Sn coils standard insulations as for NbTi wound coils are used. A special feature of the 'Wind and React' process is the brazing and insulation of splices or interconnections between adjacent layers and of the main connections to the magnet. Good splices are obtained by brazing the two Nb$_3$Sn cables

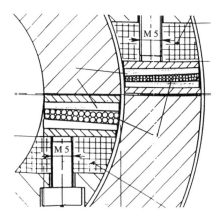

FIG. 10.20. Sandwiched NbTi–Nb$_3$Sn–NbTi end connections for Nb$_3$Sn high field magnets (courtesy: ELIN, Weiz, Austria).

in U-shaped copper forms as shown in Fig. 10. 20. The successful development and testing of the two Nb$_3$Sn wound, 1 m long, 5 cm bore mirror and full dipole magnets of the ELIN–CERN collaboration with obtained fields of 10.02 T and 9.45 T have demonstrated the validity of the 'Wind and React' process and opened the way for further developments of high field Nb$_3$Sn magnets, also on an industrial scale. Plate 3 and 4 show the two layers of a pole winding after reaction and after impregnation in a common mould.

In a graded coil design the layers are wound with cables matched to the radially decreasing magnetic induction. Superconducting dipoles in the 5–10 T range will normally have two graded layers. For the next generation of high field (dipole) magnets up to and beyond 14 T three or four layer configurations seem appropriate. A compromise will have to be found between the possibility of winding two adjacent layers with the same cable without a splice and the matching of each layer to the field. Quench propagation and quench protection schemes for high field superconducting magnets were treated in Chapter 7.

The coil interconnections and magnet-to-magnet connections should be longitudinally flexible to allow for thermal contraction, well (copper) stabilized even when placed in a low field region and supported against the acting EM forces. Cold mass supports of minimum heat load are nowadays of a standardized design, as shown in Fig. 10.21. They consist of alternating coaxial cylinders made of stainless steel and of glass fibre reinforced epoxy resin. The first are subject to tensile, the second to compressive stresses. The supports are connected to screens at intermediate temperatures between 4.2 K and ambient temperature with the aim of absorbing as much heat as possible at temperatures above 4.2 K or 1.8 K.

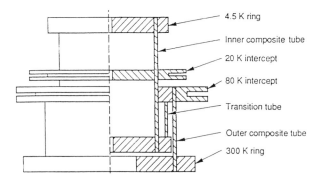

- 4.5 K ring
- Inner composite tube
- 20 K intercept
- 80 K intercept
- Transition tube
- Outer composite tube
- 300 K ring

FIG. 10.21. Cold mass support, made of concentric cylinders of stainless steel and glass-reinforced epoxy resin.

The instrumentation of superconducting high field magnets contains analogue sensors, pickup coils, voltage taps and strain gauges for temperature, electromagnetic and mechanical measurements. Analogue sensors are well described in the literature [19]. Rapid progress in fast electronics and computer technology allow the monitoring and storing of data on the quench development immediately before, during, and after the event. During the development or prototype phase a maximum of information on the behaviour of the magnet is obviously required. Hence the need for a large number of sensors. It should, however, be recalled that the presence of a large number of sensors may weaken the magnet structure, notably the insulation, and damage the excitation coils. In any superconducting magnet or string of magnets the instrumentation should be determined by safety and quench protection requirements.

10.5   Future prospects for very high field superconducting accelerator magnets

To end this chapter, let us try to make some estimates and forcasts about the future development of superconducting high field accelerator and collider magnets beyond 14 T. Prospects for these magnets have been extensively discussed at recent workshops held at the LBL, Berkeley, in 1993 and 1997 [20, 21] and at Erice, Italy in 1995 [22]. As to superconductors, consensus has been reached that cables made of Ta-doped $(Nb_3Ta)Sn$ wires, operated at 1.8 K are a realistic option—see Fig. 5.17—and that the development of other high field superconductors like Nb_3Al, APC–NbTi and Nb_3Sn conductors and of low temperature operated HTS–Bi (2223) tapes should be encouraged. Consensus has also been reached about the field levels. It is believed that superconducting dipole magnets up to 15 T could be developed within the next five years and that a further decade may be required to approach the 20 T field limit. It has also been agreed that dipole magnets up to 15 T could still be designed with cos Θ or intersecting ellipse winding configurations, even if the coils would then be exposed to higher azimuthal pressures on the horizontal midplane. To reduce these pressures winding configurations have been proposed where the main field flux lines would not cross the excitation coils. One of the proposed configuration is the flux pipe dipole and quadrupole magnets [23, 24]. The excitation coils would have a complicated shape and require about 2.8 times more superconducting cable compared to a classical $\cos \Theta$ design for the same bore field.

Figure 10.22 shows a $\cos \Theta$ winding configuration for a 15 T dipole. For the inner layer a rectangular 19.8×2.65 mm^2 cable with 30×1.325 mm Φ strands of $(Nb_3Ta)Sn$, and for the two outer layers a rectangular 18×2.58 mm^2 cable with 30×1.19 mm Φ Nb_3Sn strands, have been assumed. The coils are supposed to operate at 1.8 K; the assumed overall current densities are $j_{av} = 3.5 \times 10^8$ [A m^{-2}] in the first and $j_{av} = 4.3 \times 10^8$ [A m^{-2}] in the second and third layers at $I_0 = 20$ kA, $B_0 = 15$T. The dipole would have a double clamping mechanical structure in order to redistribute the high azimuthal stresses in the coils; preliminary computations indicate azimuthal pressures of 150 MPa $\leq -p_\Theta \leq$ 180 MPa. An interesting and promising design for future high field accelerator dipoles is the proposal of P. McIntyre [25] for a 16 T dual or twin aperture dipole. The starting idea is the subdivision of the excitation coils into several mechanically self-sustaining units or **blocks**, placed into a rigid grid structure made of inconel bars. Each coil block is wound with Nb_3Sn cables, impregnated with epoxy resin, and horizontally prestressed by an inconel spring to 15 MPa. The free volume in the spring part will be used as a superfluid helium cooling duct. The magnet and coil cross sections are shown in Figs 10.23 and 10.24. At nominal 16 T bore field the maximum horizontal coil preload amounts to $\sigma_{hm} \approx 70$ MPa. A layer of mica paper is wrapped around each block providing the mass insulation. Due to low shear resistance it is expected that the layer will also reduce training effects at the coil–inconel interface. The main parameters of the 16 T dual aperture dipole are:

Aperture	d	$=$	2.5 cm Φ
Current densities at 16 T	j_{av}	$=$	3×10^8 [A m^{-2}]
	j_{nonCu}	$=$	5.9×10^8 [A m^{-2}]
	$j_{Cuquench}$	$=$	1.15×10^9 [A m^{-2}]
Stored energy	E	$=$	2.77 [M J m^{-1}]
EM force	F_h	$=$	13 [M N m^{-1}]
Yoke dimension	56×40 cm^2		

The inconel structure is surrounded by an oval laminated yoke of magnetic steel to be clamped by an outer retaining structure. In order to obtain the required field quality in the two apertures the magnet will have to be powered by three programmed currents $I_1(t)$, $I_2(t)$, and $I_3(t)$.

It is intended to first build a 1 m long, 7.7 T dipole model wound with NbTi cable before designing and manufacturing the 1 m long, 16 T dual aperture dipole.

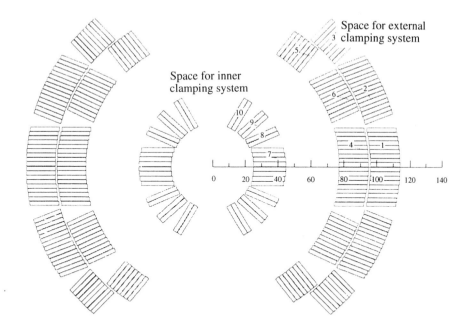

FIG. 10.22. Preliminary layout of a three layer cos Θ winding for a 15 T Nb$_3$Sn dipole cooled at 1.8 K.

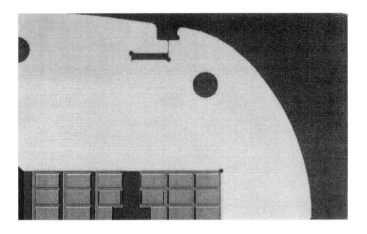

FIG. 10.23. Cross-section of a 16 T Nb$_3$Sn dual aperture dipole (courtesy: Texas A M University, College Station, USA.

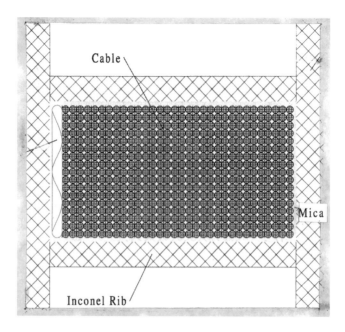

Fig. 10.24. Coil cross-section for the above dipole (courtesy: Texas A & M University, College Station, USA).

11

COOLING OF LARGE ACCELERATOR MAGNET SYSTEMS

11.1 Heat loads in large accelerator magnet systems

Heat losses and heat inleaks in superconducting magnet systems are due to:

(i) **resistive losses** at varying magnetic fields such as magnetization losses and eddy curent losses in the non-superconductive cable parts, in structural parts, current feedthrough, cable splices, end connections, etc.

(ii) **beam losses** caused by synchrotron radiation, imaging currents in the beam pipes, continuous losses in the circulating beams, and the relative inefficiency of collimators. In highly irradiated beam regions additional heat is deposited in the magnet cold masses due to absorption of secondary particles.

(iii) **heat inleaks** caused by thermal conductivity and radiation losses of magnet parts at ambient and different cryogenic temperatures.

11.2 Cooling systems

The cooling systems for the superconducting systems of the HERA, SSC, and LHC colliders will now be briefly described: the Hera cooling system has been in operation since 1987, the SSC project has regretfully been abandoned, although the cooling system had to a large extent been built and tested, and the LHC cooling system is expected to become operational around the year 2002.

The 800 GeV / 30 Gev p-e collider HERA consists of 422 ≈ 10 m long 4.6 T dipoles, and 224 ≈ 4 m long 90 T m^{-1} quadrupoles. In addition 16 superconducting cavities plus two large detector solenoids are also cooled. The HERA ring is subdivided into eight octants of 26 half cells each. A standard half cell has two dipoles and one quadrupole [1, 2]. The refrigeration scheme is shown in Fig. 11.1. The total heat losses at 4.4 K, at 40–80 K screen temperatures, and the current lead helium gas flow rates are summarized in Table 11.1.

Although two cold boxes would meet the refrigeration requirements with a fair safety margin, three interchangeable units have been installed, the third unit serving as a permanent standby. The cooling scheme is shown in Fig. 11.1. A 6.5 km long, vacuum insulated five-fold transfer line connects the three cold boxes of the central refrigerator to four feed boxes, with two precoolers each. The cold boxes supply **single phase, supercritical helium at 4.5 K** to the precooler or heat exchanger before flowing through ≈ 78 magnets of an octant. At the end the helium is expanded in a Joule–Thomson valve and returned as two-phase helium to the precooler. The evaporated helium gas flows into

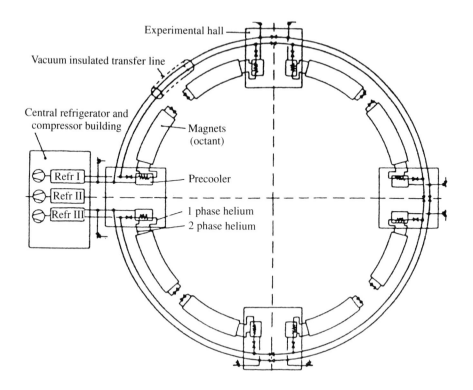

FIG. 11.1. The refrigeration system of the HERA collider.

the vapour return line of the transfer line. Continuous heat exchange by recooling takes place in the magnets, whereby the two-phase helium provides stable isothermal conditions. Each cold box has seven turbines and 14 heat exchangers. The cold boxes are fed from 14 screw compressors. In the five helium storage vessels of 4000 m³, 15 tonnes of helium at 300 K and 2 MPa or 20 bars can be stored. The system has two 10 000 *l* dewars for liquid helium plus one 15 000 *l* liquid nitrogen dewar for precooling to 80 K.

The experience with this cryogenic system is encouraging. The operation started in 1987; since 1993 the complete system has been in continuous operation at 4.3 K. The temperature variation in a dipole magnet is below 20 mK and in an octant below 100 mK. The availability of the system over two thirds of a period of operation amounted to 99% and to 98% over the entire period [2].

It may be of interest to briefly describe the cryogenic system of the abandoned 2 × 20 TeV Superconducting Super Collider Project SSC [3]. The system was conceived for the cooling of the superconducting magnets in the two vertically superimposed rings of 86 km circumference with a total of ≈ 8000 15.9 m long dipoles, 1664 5.9 m long quadrupoles, plus ≈ 1664 spool units containing correcting magnets, current and safety

Table 11.1 *Heat loads and specified refrigerator capacities of the HERA collider.*

Item	Heat load at 4.4 K [W]	Cooling gas rate at 4.4 K [g s⁻¹]	Heat load at 40–80 K [W]
Dipole	3.7		21.4
Quadrupole	8.0	0.024	38.0
One octant	907*	3.31**	2 760***
Eight octants	7256	26.5	22 080
16 s.c. cavities and 2 detectors	2000	4.4	6 000
Transfer line	1100		11 300
Total expected load	10 356	30.9	39 380
Specified total refrigeration	13 550	41.0	40 000
Total installed refrigeration	3x 6775	3x 20.5	3x 20000

* safety factor 1.5; ** safety factor 2.5; *** safety factor 1.1.

Table 11.2 *Heat loads and refrigeration capacity of the former SSC collider.*

Item	Liquid helium flow [g s⁻¹]	Heat load at 4 K [W]	Heat load at 20 K [W]	Heat load at 80 K [W]
String	5.04	924.3	1931	12 802
Sector	20.16	3697	7724	51 208
Allowance	7.84	1503	1876	13 792
Safety factor	1.25	1.25	1.5	
Sector refrigerator	35	6500	14 400	65 000

leads, vacuum and cryogenic interconnections, and other auxiliary equipment. According to Fig. 11.2 the cryogenic system is subdivided into 10 sectors with a refrigerator plant, supplying 4.3 km long upstream and downstream strings of magnets. Each string contained 24 cells of 180 m with 2 × 5 dipoles, two quadrupoles, and two spool units, plus a pool boiling recooler. The main parameters of the system are given in Table 11.2.

Single-phase helium at 4 K is pushed through a string of magnets upstream and downstream of a sector refrigerator. Small quantities of single-phase helium are tapped and expanded in pool boiling recoolers or heat exchangers, located at 180 m distance. The SSC refrigeration system, which had previously been adopted for the 1 TeV collider of Fermilab could also provide single-phase helium at lower temperatures around 3–3.5 K in order to increase, if required, excitation currents and fields in the superconducting magnets.

Finally in this chapter the challenging and complex cooling system of the 27 km circumference 2 × 7 TeV LHC collider of CERN will be described. The complexity of the

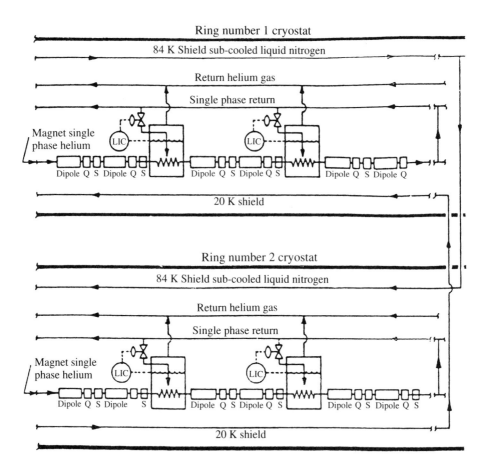

FIG. 11.2. The refrigeration system of the SSC collider.

system is primarily due to the requirement that the magnets be cooled with **superfluid helium at 1.8–1.9 K**. The planned system represents a considerable step forward with respect to the first large superfluid helium cooling system, which had been developed for the Tore-Supra magnet described in [4]. The basic requirements on the LHC cooling system are as follows.

The total cold mass of the LHC magnets of 31 000 tonnes should be cooled down to 1.8–1.9 K within 25 days.

The system should be able to cope with quenches and resistive transitions which may occasionally occur, to safely evacuate the released heat, and to withstand the high pressure waves in the helium vessels attaining 2 MPa or 20 bars in case of a quench.

The quench propagation to adjacent magnets must be limited and the down-time

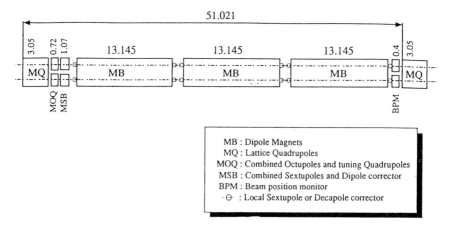

FIG. 11.3. Layout of a LHC half cell of superconducting magnets.

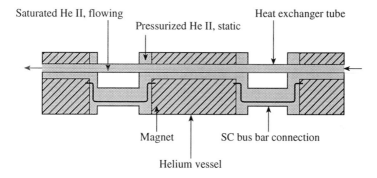

FIG. 11.4. Principle of the LHC magnet cooling with superfluid helium.

following quenching of a 50–100 m long string of magnets should not exceed a few hours.

The cooling system must respect the constraints of the already existing 27 km tunnel of the LEP collider, such as its eightfold symmetry. Due to the 1.41% tunnel inclination and the elevation variations up to 120 m, hydrostatic heads would be created in a two-phase liquid plus vapour flow. All fluids must over large distances be transported as single-phase or subcooled liquid, superheated vapour, or supercritical helium. However, local two-phase circulation of saturated or boiling helium can be tolerated within a specified range of helium vapour (or quality factor).

The LHC cryogenic system has three cooling temperature levels:

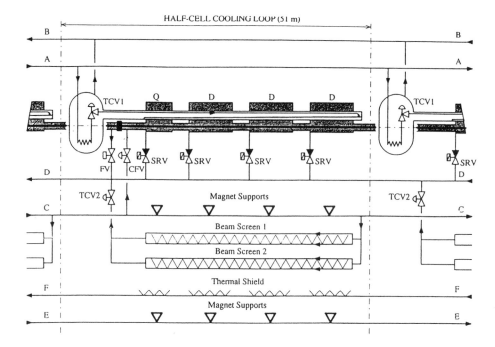

FIG. 11.5. Cryogenic flow chart for a LHC half cell.

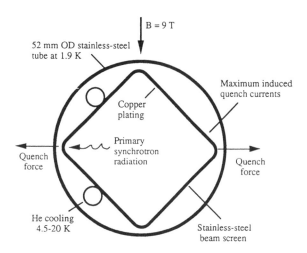

FIG. 11.6. Cross-section of the cold bore stainless steel tube.

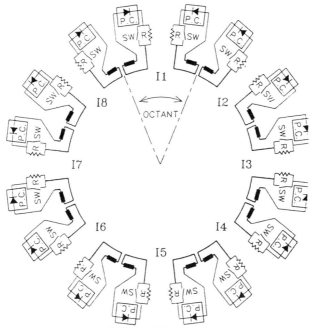

I Intersection point   P.C. Power converter
SW Switch   R Dumping resistor

FIG. 11.7. Powering scheme for the LHC magnets.

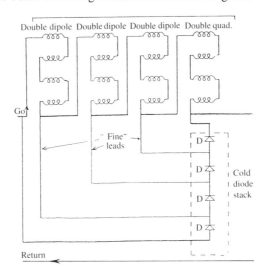

FIG. 11.8. Quench protection scheme for a half cell of LHC magnets.

Table 11.3 *Installed cryogenic power requirements per octant of the LHC*

Required power [kW]	at 50–75 K	at 4.5–20 K	at 4.7 K	at 1.9 K	liquid He [g s^{-1}]
	30–32	5.0	0.1–1.1	20–24	31–53

- the outer shielding level at temperatures between 50 and 75 K;
- the heat interception or inner screening level at temperatures of 4.5–20 K;
- the magnet cold mass temperature of 1.8–1.9 K.

Based on the calculated heat loads and losses, the required installed refrigeration power per machine octant is given in Table 11.3.

The LHC cooling system extends over a 24 km circumference, out of which ≈ 20 km are occupied by eight regular arcs of standard FODO cells. Each arc has 24 cells of ≈ 100 m length. The basic cooling loop extends over a 50 m long half cell and contains three main ≈ 14 m long twin aperture dipoles, one main ≈ 3 m long twin aperture quadrupole, and a number of correcting magnets. The principle of the LHC magnet cooling is shown in Fig. 11.4. The magnet cold masses are immersed in **static, pressurized 1.9 K superfluid helium at 1 bar**, which is cooled by heat exchange with **saturated superfluid helium at 16 mbar**. The two systems are thus hydraulically insulated. The cooling loop of a half cell is shown in Fig. 11.5. During initial cool-down the magnet cold masses are cooled to 4.5 K by liquid helium flowing through lines C and D. Subcooled superfluid helium at 2.2 K and 1 bar is supplied by line A, expanded in the J–T valve and returned via the 1.8–1.9 K, 16 mbar pumping line B. The beam screens (1) and (2) are nonisothermally cooled at 4.5–20 K , the thermal shield being at 50–75 K. The validity of the cooling system has been confirmed on a superfluid helium test cryoloop and more recently by cooling down and operating a half cell at CERN. The LHC cooling system thus provides quasi-isothermal cooling of the magnet cold masses at 1.8–1.9 K. The temperature increase in the saturated superfluid helium, transporting the heat losses over 1670 m of a half-octant to the refrigerator, is less than 0.1 K. The cross-section of the magnet cold bore with a 52 mm Φ stainless steel tube, cooled at 1.9 K, and the inner square stainless steel screen, cooled at 4.5 K–20 K is shown in Fig. 11.6.

Closely related to LHC cryogenic layout are the powering and quench protection schemes. The magnets are powered from 16 power converters, two per octant, as shown in Fig. 11.7. The advantages of this scheme are the reduced steady state power consumption, the reduced ground voltage due to galvanic isolation and to the 16-fold earthing, and the need to discharge only one sixteenth of all ring magnets in the case of a quench. The main disadvantage is the required high precision (10^{-5}) tracking of the power converters such that all magnets appear to be powered in series. The quench protection scheme is shown in Fig. 11.8. It must prevent the magnets of a half-octant from discharging their energy into the quenching unit and destroying it. To this end the protection relies on three elements: the quench heaters, the bypassing cold diodes, and the rapid de-excitation of the series connected unquenched magnets. The radiation resistant cold diodes are placed in the superfluid helium vessel. If a magnet, belonging to a half cell, quenches, the remaining magnets of this half cell will also quench by activating their heaters and discharging through the four series connected diodes, each diode connected in parallel to the dipoles and the quadrupole of that half cell. Fast de-excitation of a half octant with a time costant of $\approx$ 100 ms is obtained by rapid switching of high power dumping resistors.

HIGH FIELD SOLENOIDS AND DETECTOR MAGNETS

12.1 Introduction

High field solenoidal magnets or solenoids belong to the second large group of super-conducting high field magnets to be treated in this book. In fact, solenoids were the first devices to be wound and tested since the emergence of technical superconductors, which started with the develop - ment of Nb_3Sn wires by Kunzler *et al.* in 1951 [1]. Solenoids are still efficient devices for the testing of technical superconductors and for the verification of specific coil winding techniques.

In addition to this vast group of test coils with rather modest requirements on field precision, two further groups will be adressed: the rapidly expanding family of **NMR** (nuclear magnetic resonance) and, for medical purposes, **MRI** (magnetic resonance imaging) solenoids and large size solenoids for **detectors** in high energy physics experiments. NMR and MRI solenoids are characterized by the required high field uniformity of $\frac{\Delta B}{B} \leq 10^{-8}$ within a specified volume. NMR devices have recently been built for fields beyond $B_0 \geq 20$ T in volumes of ~ 1 m^3. In detector solenoids for high energy physics experiments with particle accelerators and colliders the magnetic induction should be known in the entire volume with a precision of 10^{-3}–10^{-4} and to better than 10^{-4} within a central volume of the solenoid. Due to the size of these coils the main problems to be solved are the mechanical design, cooling and matching of a large number of complex elements cohabitating in such a detector. The fourth group of still larger, very high field solenoids and toroids for future energy production fusion generators will not be adressed, since they are considered to be beyond the scope of this book.

12.2 Computation of magnetic fields in solenoids

Expressions for the magnetic field and induction in solenoidal winding configurations can be derived from a circular current loop, shown in Fig. 12.1. The magnetic induction $B_z = \mu_0 H_z$ at point $P(z)$ along the z-axis due to the loop current I is [2]

$$B_z = \frac{\mu_0 I}{2a} \sin^3 \alpha. \tag{12.1}$$

For $\alpha = \frac{\pi}{2}$ one obtains B_{z0} in the loop centre:

$$B_{z0} = \frac{\mu_0 I}{2a}. \tag{12.2}$$

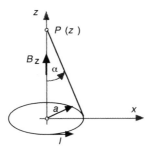

FIG. 12.1. Magnetic field of a circular current loop.

For the circular helix shown in Fig. 12.2 of radius a, step p, with N turns, and

$$b = \frac{Np}{2} - c, \tag{12.3a}$$

one obtains for B_z at the point $P(x, y, z)$

$$B_z = \frac{\mu_0 I}{2p} \left\{ \frac{N\frac{p}{2} + b}{\left[a^2 + \left(\frac{Np}{2} + b \right)^2 \right]^{0.5}} + \frac{\frac{Np}{2} - b}{\left[a^2 + \left(\frac{Np}{2} - b \right)^2 \right]^{0.5}} \right\}. \tag{12.3b}$$

For a thin solenoid of length l, according to Fig. 12.3, one obtains for B_x at the point P_x along the axis:

$$B_x = \frac{\mu_0 n I}{2l} (\cos \alpha_1 - \cos \alpha_2). \tag{12.4}$$

For $\alpha_1 < 0$, $\alpha_2 < 0$, $P(x)$ is outside the solenoid and for $\alpha_1' > 0$, $\alpha_2' < 0$, $P'(x)$ is within the solenoid.

For a thick solenoid of length l and winding radii r_2, r_1 shown in Fig. 12.4 one obtains for B_x at points P_x and P_x' assuming a uniform current density of

$$j = \frac{nI}{l(r_2 - r_1)}, \tag{12.5}$$

$$B_x = \frac{\mu_0 j}{2} \left[a' \ln \frac{\text{tg} \left(\frac{\alpha_2}{2} + \frac{\pi}{4} \right)}{\text{tg} \left(\frac{\alpha_1}{2} + \frac{\pi}{4} \right)} \pm b' \ln \frac{\text{tg} \left(\frac{\beta_2}{2} + \frac{\pi}{4} \right)}{\text{tg} \left(\frac{\beta_1}{2} + \frac{\pi}{4} \right)} \right]. \tag{12.6}$$

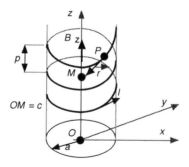

FIG. 12.2. Magnetic field of a circular helix.

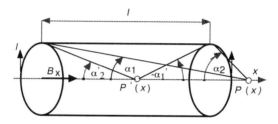

FIG. 12.3. Magnetic field of a thin solenoid.

The sign (+) stands for $P(x)$ inside; the sign (−) for $P'(x)$ outside the solenoid. For the off-axis components B_x and B_ρ one can expand the magnetic scalar potential in the central plane and along the x-axis:

$$V^*(\rho, x) = V^*(0, x)\left[1 + \sum_{n=1}^{\infty} 1 + \left[\frac{(-1)^n}{(n!)^2}\left(\frac{\rho}{2}\right)^2\left(\frac{\delta}{\delta x}\right)^{2n}\right]\right]. \quad (12.7)$$

With

$$B_x(\rho, x) = -\mu_0\frac{\delta V^*}{\delta x} \quad (12.8)$$

$$B_\rho(\rho, x) = -\mu_0\frac{\delta V^*}{\delta \rho} \quad (12.9)$$

one obtains:

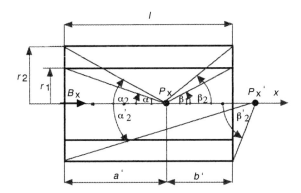

FIG. 12.4. Magnetic field of a thick solenoid.

$$B_x(\rho, x) = \sum_{n=1}^{\infty} \frac{(-1)^n}{(n!)^2} \left(\frac{\rho}{2}\right)^{2n} \left(\frac{\delta}{\delta x}\right)^{2n} [B(0, x)] \tag{12.10}$$

$$B_\rho(\rho, x) = \sum_{n=1}^{\infty} \frac{(-1)^n}{n!(n-1)!} \left(\frac{\rho}{2}\right)^{2n-1} \left(\frac{\delta}{\delta x}\right)^{2n-1} [B(0, x)]. \tag{12.11}$$

The axial field along either the x-axis or the central plane according to eqn (12.11) can also be written in cylindrical coordinates [3, 4] as

$$B_x(x, 0) = B_0 \left[1 + E_2 \left(\frac{x}{r_1}\right)^2 + E_4 \left(\frac{x}{r_1}\right)^4 + \ldots + E_{2n} \left(\frac{x}{r_1}\right)^{2n}\right] \tag{12.12}$$

$$B_x \left(\rho, \theta = \frac{\pi}{2}\right) = B_0 \left[1 - \frac{1}{2}E_2 \left(\frac{\rho}{r_1}\right)^2 + \frac{3}{8}E_4 \left(\frac{\rho}{r_1}\right)^4 - \frac{5}{16}E_6 \left(\frac{\rho}{r_1}\right)^6 \pm \ldots\right]. \tag{12.13}$$

B_0 is the on-axis field in the solenoid centre and

$$E_{2n} = \left| \frac{1}{B_0} \frac{1}{(2n)!} \frac{\delta^{2n}}{\delta x^{2n}} B_x(x, 0) \right|_{x=0} \tag{12.14}$$

One of the problems to be adressed and solved concerns optimized solenoid geometries yielding the required field uniformity within a specified volume at a minimum number of concentric layers. Entire books had been written to achieve this aim such as those by B. Montgomery and Chari and Silvester [3, 5], and analytical and graphical solutions have been elaborated for field uniformities around $\frac{\Delta B}{B_0} \approx 10^{-4}$, also called sixth-order coils. It should, however, be borne in mind that these calculations were performed 15–25 years ago when powerful computer codes for magnetic field computations were not available.

Today one can use the calculated geometries as starting data for inverse computer codes for further optimization respecting other input conditions and constraints: the maximum number of individual coils and coil layers, the overall dimensions of the solenoid, the superconductor or superconducting cable load lines, the nominal current (s), the required field uniformity in a given volume, and the minimum volume of superconductor (s). The merits of active or passive screening by superconductive coils or iron screens should also be considered.

12.3   Calculation of mechanical forces and stresses in solenoids

The magnetic induction $B(z, \rho)$ in a long vertical solenoid exhibits a dominant axial component $B_z(z, \rho)$, which will in the coil give rise to a radial force $F_\rho(z, \rho)$. F_ρ is directed outwards if B_z is oriented in the direction of the bore field. In the coil region of the return flux with $-B(z, \rho)$, $-F_\rho$ is directed inwards. The radial component of the magnetic induction $B_\rho(z, \rho)$, increasing towards the ends of the solenoid, gives rise to axial end forces $-F_z(z, \rho)$, which always compress the coil ends. For a thin solenoid of inner coil radius r_1 and an average induction in the winding B_{wzav}, one obtains for the tangential stress in the winding σ_t [Nm^{-2}] with t the coil thickness:

$$\sigma_t = B_{wzav} \frac{Ir_1}{lt} = 0.5 B_{wzav} j r_1. \tag{12.15}$$

For the axial compressive stress $-\sigma_z$ [Nm^{-2}] one obtains:

$$\sigma_z = -B_{w\rho av} j \Delta l. \tag{12.16}$$

As shown in Fig. 12.5 an equilibrium exists in a coil element between the EM body force $f_b = B_w j \rho d\rho d\theta dz$, the radial inward component $f_\rho = -\sigma_t \Delta\rho \Delta\theta \Delta z$ due to the circumferential tensile force $T = f_t$ and $f_p = \frac{d}{d\rho}(\rho\sigma_\rho)\Delta\rho\Delta\theta\Delta z$, and the difference of the radial pressure force over $d\rho$, such that

$$f_b + f_\rho + f_p = 0 \tag{12.17}$$

or

$$-\sigma_t + \frac{d}{d\rho}(\rho\sigma_\rho) = -B_w j \rho. \tag{12.18}$$

Introducing the Poisson coefficient v and the (isotropic) elasticity modulus E [6], σ_ρ, σ_t can be expressed in terms of a single variable, the radial displacement u:

$$\sigma_\rho = \frac{E}{1 - v^2}(\varepsilon_\rho + v\varepsilon_t) = \frac{E}{1 - v^2}\left(\frac{du}{d\rho} + v\frac{u}{\rho}\right) \tag{12.19}$$

$$\sigma_t = \frac{E}{1 - v^2}(\varepsilon_t + v\varepsilon_\rho) = \frac{E}{1 - v^2}\left(\frac{u}{\rho} + v\frac{du}{d\rho}\right) \tag{12.20}$$

Introducing the last two equations into eqn (12.18) one obtains:

$$\frac{d}{d\rho}\left(\rho\frac{du}{d\rho}\right) - \frac{u}{\rho} = -\frac{1-v^2}{E}\rho B_w(\rho)j(\rho). \tag{12.21}$$

Equation (12.21) can be solved for solenoids of finite length l if one assumes, to a fair approximation, a linear decrease of B_z from B_{z1} at $\rho = r_1$ to B_{z2} at $\rho = r_2$:

$$B_z = \frac{(r_2 - \rho)B_{z1} + (\rho - r_1)B_{z2}}{r_2 - r_1} \tag{12.22}$$

Normalizing and introducing the following terms,

$$e = \frac{\rho}{r_1}; \ \alpha = \frac{r_2}{r_1} \tag{12.23}$$

$$w = \frac{uE}{r_1(1-v^2)} \tag{12.24}$$

$$K = \frac{(\alpha B_{z1} - B_{z2})jr_1}{\alpha - 1} \tag{12.25}$$

$$M = \frac{(B_{z1} - B_{z2})jr_1}{\alpha - 1} \tag{12.26}$$

eqn (12.22) becomes

$$\frac{1}{e}\frac{d}{de}\left(e\frac{dw}{de}\right) - \frac{w}{e^2} = -K + Me \tag{12.27a}$$

with the solution:

$$w = Ce + \frac{D}{e} - K\frac{e^2}{3} + M\frac{e^3}{8}. \tag{12.27b}$$

The coefficients C, D are determined from the boundary conditions at r_1 and r_2, where in most cases $\sigma_\rho = 0$. The expressions for σ_t and σ_ρ are:

$$\sigma_t = \frac{K(2+v)}{3(\alpha+1)}\left[\alpha^2 + \alpha + 1 + \frac{\alpha^2}{e^2} - e\frac{(1+2v)(\alpha+1)}{2+v}\right] \tag{12.28}$$

$$- M\frac{2+v}{3(\alpha+1)}\left[\alpha^2 + 1 + \frac{\alpha^2}{e^2} - e^2\frac{(1+3v)}{3+v}\right]$$

$$\sigma_\rho = K\frac{(2+v)}{3(\alpha+1)}\left[\alpha^2 + \alpha + 1 + \frac{\alpha^2}{e^2} - (\alpha+1)e\right] \tag{12.29}$$

$$- M\frac{3+v}{8}\left[\alpha^2 + 1 - \frac{\alpha^2}{e^2} - e^2\right]$$

Figure 12.6 shows $\sigma_t(e)$ and $\sigma_\rho(e)$ normalized with respect to the magnetic pressure $P = \frac{B^2}{2\mu_0}$ for a solenoid with $\alpha = 1.8$. σ_t' represents the rather unrealistic case for the coil turns acting independently of each other.

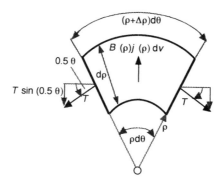

FIG. 12.5. Force equilibrium in an element of a solenoid.

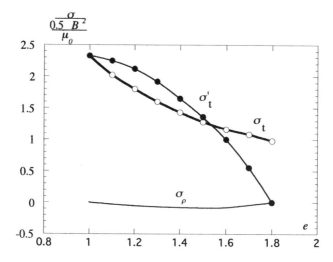

FIG. 12.6. Diagram of normalized stresses σ_t and σ_ρ.

Stress containment in superconducting high field solenoids follows the principles applied to transverse field windings, described in Chapter 10. The radial and circumferential stresses σ_ρ and σ_t should if possibe be compressive. To this end the coils should be radially compressed by welded or heat shrunk stainless steel or Al cylinders or by wrapped and spot-welded tapes or wires around half shells. Expressions for thick wall cylinders and for composite cylinders can be applied [6], whereby the inner cylinder represents the coil limited by the radii r_2 and r_1 and the retaining system, and the outer jacket by the radii r_2 and r_3, the respective elasticity moduli being E_{12} and E_{23}. The

Table 12.1

Particle	Spin I	$\frac{\gamma}{2\pi}$ [MHz T^{-1}]
Electron e$^-$	$\frac{1}{2}$	$2.8025.10^4$
Proton p$^+$	$\frac{1}{2}$	42.567
3_{He}	$\frac{1}{2}$	32.3427
27_{Al}	$\frac{5}{2}$	11.093
Deuteron D	1	6.535

applied coil prestress should be such that

$$|-\sigma_t| > |\sigma_{t\,coil}| \text{ at } \rho = r_1; \ \sigma_\rho \approx 0 \text{ at } \rho = r_1. \tag{12.30}$$

In high field and high j_c solenoids of limited radial extension it may not always be possible to satisfy conditions (12.30). A limitation can then be imposed on the maximum tensile strain, notably in coils wound with prereacted Nb$_3$Sn—see also Section 5.5.6.

12.4 Basic principles of nuclear magnetic resonance

Before addressing the problems related to the design of very high field NMR spectrometer magnets the basic principles of nuclear magnetic resonance will briefly be given. The phenomenon was discovered by E. Bloch [7] and has since been applied to high precision measurements of magnetic fields and for medical diagnostics. Basically elementary particles or molecules like protons, deuterons, electrons, or muons with a spin I and angular momentum L will exhibit a **magnetic moment** μ:

$$\mu = \gamma \hbar I = \gamma L \tag{12.31}$$

with $\hbar = \frac{h}{2\pi}$, h being Planck's constant and γ [MHz T^{-1}] the gyromagnetic ratio of the particle. In a magnetic field of induction B [T] a particle will exhibit a precessional motion with a precession or Larmor frequency

$$v_L = \frac{\omega_L}{2\pi} = \frac{\gamma}{2\pi}|B|. \tag{12.32}$$

Equation (12.32) is fundamental for NMR field measurements as it links the magnetic induction B to the frequency v_L which can be measured with high precision. Table 12.1 gives the properties for some elementary particles and molecules.

Table 12.1 suggests that electrons are well suited for measurements of low inductions around 0.01 T and protons for high inductions between 0.5 and 10 T. Very high magnetic inductions of $B > 20$ T are of considerable interest to medical applications as they are proportional to the size and resolution of the molecules to be investigated in human and animal bodies.

Table 12.2 *Main parameters of the 900 MHz 21.1 T*
high field solenoid

Bore field	21.1 T
Operating temperature	1.8 K
Clear bore	0.11 m
Magnet bore	0.142 m
Magnet outer diameter	0.77 m
Magnet length	1.44 m
Weight of Nb_3Sn superconductor	700 kg
Weight of Nb Ti superconductor	1050 kg
Weight of cold mass	2970 kg
Nominal current	350 A
Magnet inductance	414 H
Stored energy	25.4 MJ

12.5   Design and manufacture of high field NMR solenoids

Superconducting NMR spectrometer solenoids cover a vast field range of $1\,T \leq B \leq 25$ T. Their design and manufacturing problems are closely related to the solenoid geometry, the magnetic induction, average current density in the coils and to the resulting EM forces, stresses, and strains. To show how these problems have been adressed and solved the high field NMR magnet development at the US National High Magnetic Field Laboratory (NHMFL) will be briefly described [8, 9, 10]. The programme has three phases: the development of superconducting NMR solenoids for 750 MHz (proton) resonance, followed by a 900 MHz 21.1 T unit. During the third phase an insert wound with a high B_{c2} superconductor will increase the field to $B_0 = 25$ T, which corresponds to a frequency of 1066 MHz. The 750 MHz solenoid was completed in 1993, and the technology for the 900 MHz solenoid was developed and successfully applied to a 20.4 T test coil assembly.

Figure 12.7 shows the cross-section of the 900 MHz 21.1 T solenoid; the magnet consists of nine main coils. The five inner coils are wound with Nb_3Sn cable, the four outer ones with NbTi cable. All coils are graded with the operating or nominal current densities j_{av} matched to the highest local inductions. The main parameters of the 900 MHz 21.1 T NMR solenoid are given in Table 12.2; The eight-strand Nb_3Sn cable with a glass-web insulation and the interlayer quartz-braid insulation is shown in Fig. 12.8. To achieve the highest possible j_{av}, the Nb_3Sn and NbTi coils are cooled by superfluid He II at 1.8 K. The gain in $j_{c\,non\,Cu}$ obtained in Nb_3Sn superconductors and the influence of mechanical strain are shown in Fig. 5.17b. Figure 12.9 shows the calculated 10^{-8}–10^{-9} field uniformity regions in the solenoid.

Technological and manufacturing aspects were adressed by building a 20.4 T bore field, 36 cm long model with a 9.2 cm bore, shown in Fig. 12.10 [10]. The model served for optimizing the reaction process after winding, including low temperature pre-anealing and anealing of the internal Sn, the Nb_3Sn cable, and of the coil impregnation.

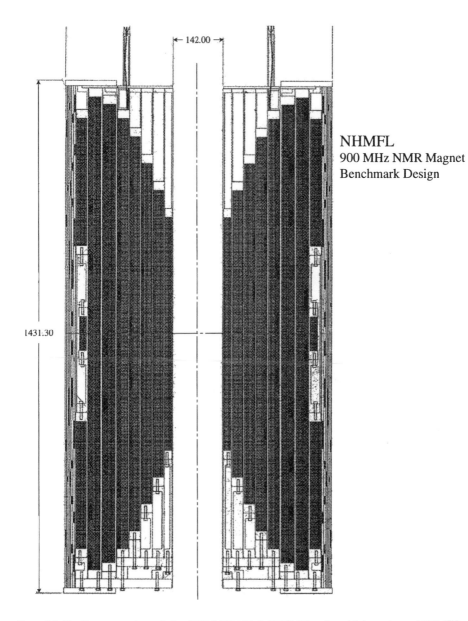

NHMFL
900 MHz NMR Magnet
Benchmark Design

FIG. 12.7. Cross-section of the 900 MHz 21.1 T NMR solenoid (courtesy: NHMFL, Florida State University, Tallahassee, USA).

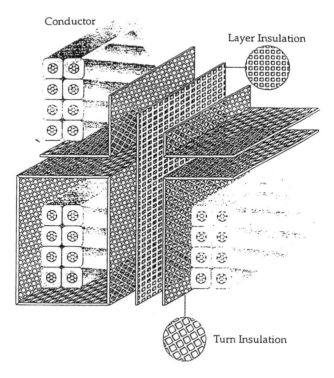

FIG. 12.8. Conductor and layer insulation of the 900 MHz 21.1 T NMR solenoid. (Courtesy: NHMFL, Florida State University, Tallahassee, USA.)

Bonding of the winding to the coil form was avoided and the coil terminals and flanges were just locally stuck to the coil end surface. A mica paper layer was placed in between the inner cylinder and the coil. Due to the low shear yielding stress of the mica paper of ~ 0.03 MPa the epoxy-impregnated cylinder plus coil assembly showed no training as observed on samples without the mica paper. The terminals of such free-moving coils must accommodate for the brittleness of the Nb_3Sn cables. One solution consists in providing flanges holding the terminals, which are then in turn bonded to the coil. A flexible strain relieving section of conductor must be forseen between the free coil body and the terminals. The model coil reached a critical current performance at 20.4 T bore field and behaved essentially as a free standing coil.

12.6 Large superconducting solenoids and toroids for high energy physics particle detectors

For high energy particle colliders a new type of experimental area has emerged with huge, densely packed multipurpose detector elements, requiring large volumes of high

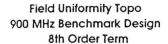

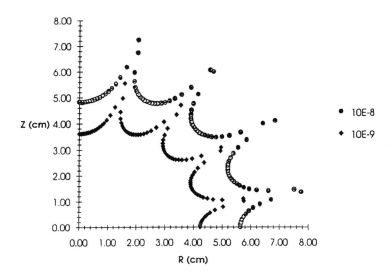

FIG. 12.9. Computed $10^{-8} - 10^{-9}$ field uniformity region.

magnetic fields, produced by superconducting 'thin wall' solenoids and in certain cases also by superconducting toroids. The development of large detector solenoids extends from the early 'time projection chamber' (TPC) solenoid of LBL [11], followed by the 'Cello' and 'Clio' solenoids [12, 13], the CDF detector of Fermilab [14], the 'Topaz', 'Venus', and AME detectors of KEK, Japan, [15, 16] to the large solenoids DELPHY and ALEPH for the LEP collider of CERN [17, 18]. The series will continue with the two approved large detectors CMS and ATLAS for the future LHC collider of CERN [18, 19, 20].

These multipurpose detectors consist essentially of three major parts, nested within each other; the central detector or tracker, close to the beam intersecting point, the calorimeters, and the external muon detecting system, as shown in Plate 5, representing the future CMS detector for the LHC collider. Particle momentum measurements are performed in the tracker and the muon detection parts where high magnetic fields are required in order to obtain a high momentum resolution. The general requirements on the magnet system are as follows.

(i) It should be well integrated into the detector space, occupied by the coils; the space for the mechanical and cryogenic parts should be small. In order to reduce the radiation length 'light' structural materials like aluminium are prefered.

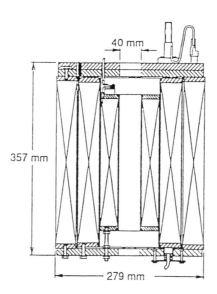

FIG. 12.10. Cross-section of the 20.4 T bore field coil model. (Courtesy: NHMFL, Florida State University, Tallahassee, USA.)

(ii) The particles to be measured should if possible cross the field lines at right angles.

(iii) In a completely assembled and installed detector the superconducting solenoids or/ and toroids will hardly be accessible and any failure or repair would require a considerable time. Reliable and safe operation of all detector elements is therefore required.

Results and experience with the operation of large volume detector solenoids are mentioned in [18, 20]. The demonstrated high quality and reliability leads us to formulate a number of basic design features like the utilization of monolithic or cabled, Al-stabilized conductors, adoption of the 'inner coil winding into an outer cylinder' technique, vacuum epoxy resin impregnation once the coils are forced into the cylinder, indirect cooling and suppression of any helium vessel.

Let us describe the ALEPH detector solenoid, built in 1987 for the LEP collider of CERN and the planned solenoidal and toroidal coils for the future LHC detectors CMS and ATLAS. The ALEPH solenoid, shown in Plate 6, is surrounded by a huge flux return iron yoke and provides a field of 1.5 T in a volume of 5 m diameter and 6.3 m length. For the central TPC detector a 10^{-4} field uniformity is required within a volume of 4.4 m length and 3.6 m diameter. This demand imposes strict tolerances on the winding geometry and requires correcting coils to accomodate for the variations in the iron permeabilty. The nominal current amounts to 5 kA, the stored energy to 136 MJ, and the total weight is 65 tonnes. Cooling is achieved by circulating liquid helium in

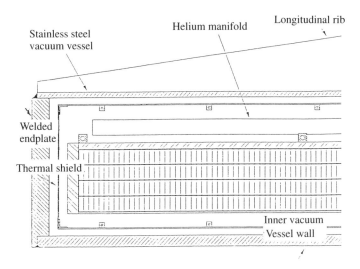

FIG. 12.11. Cross-section of the CMS four layer solenoidal winding.

pipes in good thermal contact with the Al-stabilized superconducting cable. The single layer coil is wound into the retaining cylinder, while the thin compensating coils are wound on the outer cylinder surface. The cooling is based on a thermosyphon process: liquid helium is stored above the solenoid and connected through an insulated line to a manifold, supplying the parallel connected cooling pipes at the coil bottom. The two-phase helium in these pipes, produced by heat losses, is returned to the reservoir by the difference between the upper and lower column fluid density. The process is self-regulating, as the two-phase helium flow increases with higher heat loads. The main 6.6×1.4 mm^2 Al-stabilized conductor consists of 16 strands of 0.8 mm diameter. The overall aluminium cross-section is 3.5×13.6 mm^2. At 2 T, 4.6 K the critical current amounts to $I_c = 12$ kA. For the coil winding the outer Al-alloy cylinder was placed vertically on a turntable and a cylindrical cage displaced vertically within the cylinder. The conductor was fed from a spool and formed, cleaned, sandblasted, bent, insulated, and guided along a helix before reaching its final position. A ring of rollers and pressing pads, fixed to the bottom of the cage, positioned and pressed the conductor into the cylinder by applying a controlled radial and axial prestress to the winding. During this operation the conductor was insulated with two overlapped layers of glass tape and the cylinder with four layers of glass tape strips. After winding of the correcting coils on the cylinder outer face, the assembly was placed into an *in situ* made impregnation mould and vacuum impregnated with epoxy resin.

The cryostat, clad with superinsulation on its inside, is a vacuum tank consisting of two cylindrical Al-alloy shells, connected at both ends by bolted flanges to an annular stainless steel endplate. Within the cryostat the 25 tonnes coil is supported by two pairs

Table 12.3 *Parameters of the CMS - detector solenoid*

Central induction	4.0 T
Bore free diameter	5.9 m
Magnetic length	14.0 m
Nominal current	20.0 kA
Stored energy	2850 MJ
Inductance	14.25 H
Total conductor length	48 km
Number of layers	4
Turns per layer	146
Axial midplane force	7400 tonnes
Peak axial compression	9900 tonnes
Total cryostat mass	240 tonnes
Total coil mass	210 tonnes
Content of Nb Ti	1.78%
Content of Cu	2.49%
Content of pure Al stabilizer	29.74%
Content of Al alloy	62.75%

of legs made of glass reinforced epoxy and positioned by a set of axial and radial rods. The solenoid was excited to a nominal current of 5 kA within 50 minutes.

Quenching the magnet at I_n resulted in a temperature increase of $\Delta T = 52$ K and a helium pressure rise of $\delta p = 4.8$ bar.

The experience gained with the ALEPH solenoid was incorporated into the conceptual design of the large solenoid for the CMS detector of the LHC collider [19, 20]. The CMS detector of 22 m length, 14 m diameter and a total weight of $\sim 12\,000$ tonnes is shown in Plate 5. Guided by the principle of easy access to the inner parts, the detector is conceived of seven movable subunits: two end caps and five so-called barrels, each consisting of five 2.6 m long rings. The central barrel ring supports the superconducting solenoid. The two end caps consist of three separable iron disks of 0.6 m thick iron blocks, to be assembled and welded *in situ*. The 4 T solenoid is surrounded by a heavy iron yoke, made of several layers. The cross-section of the four layer solenoid winding is shown in Fig. 12.11. A NbTi cable with 30×1.6 mm diameter strands of 2.2×26 mm^2 cross-section, embedded in a 17×40 mm^2 high purity Al stabilizer, which is in turn surrounded by an Al alloy of 22.3×62 mm^2 for mechanical strength, was used. The conductor is designed for a limiting radial strain of 0.15%, which corresponds to a radial pressure of 7 MPa. The maximum hoop stresses in the winding are below 100 MPa. Due to manufacturing and transport limitations the coil is subdivided into four parts of equal length. No correcting coils are required, as no special requirements are imposed on the magnetic field uniformity; a precise field map within the entire volume of the solenoid is, however, required. The main coil parameters are summarized in Table 12.3.

Table 12.4 *Main parameters of the ATLAS detector central solenoid*

Central field	2.0 T
Coil diameter	2.44 m
Coil length	5.3 m
Nominal current	8.0 kA
Conductor size	28.31 mm^2
Stored energy	40 MJ
Cold mass weight	5 tonnes

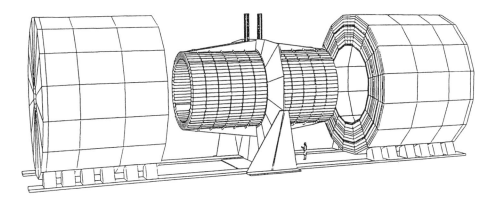

FIG. 12.12. Expanded view of the ATLAS detector.

The refrigeration system of the solenoid is again based on indirect cooling by the thermosyphon effect. The main cooling system consists of a set of Al pipes, wound around and welded to the outside cylinder of the Al alloy and connected to insulated manifolds. The solenoid has its own refrigerator for cool-down and normal operation. The required cooling capacity is 900 W at 4.5 K and 3400 W at 60 K.

The cross-section of the superconducting windings of the second LHC detector ATLAS is shown in Plate 8 [21, 22]. The detector has three systems of coils, the central thin solenoid, the eight barrel coils or toroids, and the two end cap toroids with eight coils each. The central solenoid is located inside the vacuum chamber of the calorimeter, such that valuable space and material for a separate cryostat could be saved. The magnetic field amounts to 2 T. The main parameters of the solenoid are summarized in Table 12.4.

The toroidal magnet system consists of a 26 m long barrel of 9.4 m inner and 19.5 m outer diameter. Each of the eight pancakes of the toroid has its own cryostat, while the eight coils of the endcap toroids are located in a common cryostat. An expanded view of the central barrel toroids and of one end cap toroid is given in Fig. 12.12. The large inward directed forces of the central toroidal coils are sustained by a central plate, to which are fixed the four pancakes of each coil. The torus is supported by four pillars, anchored to the floor and to the coils. All coils and pancakes of the three magnet systems have indirect cooling. The complex mechanical structure which must contain the forces within and between the toroidal coils has been determined by the ANSYS and CASTEM codes in order to ensure stability and small enough deformations under different loading conditions. The total cooling power required for the three magnet systems of the ATLAS detector is 2500 W at 4.5 K and 16500 W at 40 K.

MAGNETIC MEASUREMENTS OF SUPERCONDUCTING MAGNETS

13.1 Introduction

The history of magnetic measurements coincides with the discovery of electromagnetism during the first half of the nineteenth century, notably of the laws related to electromagnetic induction by M. Faraday in 1831. If the measuring methods remained to a large extent unchanged, the measuring equipment has been steadily improved due to constant progress in new and better materials, precision mechanics, fast electronics, and computers.

In this chapter we shall concentrate on high precision magnetic field (or induction) measurements of superconducting accelerator and large detector magnets, on the comparison of different measuring methods, and on the requirements the measuring equipment should satisfy. Error sources and their compensation will also be discussed. Although important to the magnet design, transient magnetic field measurements will not be dealt with.

Superconducting accelerator magnets are characterized by a small cross-section versus length ratio and can be considered as 2D devices. The predominantly cylindrical apertures speak in favour of specific measuring devices like rotating coils. The magnetic induction should be measured with a precision better than 10^{-4}; close and accurate relations should be established between the magnetic field axis, the symmetry planes, and the mechanical structure of the magnet. To this end the geometrical parameters of the magnet should be known with a precision of a few 0.01 mm and the angles with a few 0.01 mrad.

In large volume superconducting detector magnets of solenoidal or toroidal type the required measuring precision is by about an order of magnitude lower than for accelerator magnets. The magnetic field is usually determined by point-to-point measuremnts in the required detector volume [1].

13.2 The main magnetic field measuring methods

Accurate magnetic field measuring methods can be divided into three groups: (i) fluxmeter methods with rotating coil systems, (ii) Hall-probe measurements, and (iii) NMR (nuclear magnetic resonance) devices for high precision field measurements and calibration purposes. Fig. 13.1 gives the accuracies in ppm which can be obtained by each method as a function of the magnetic induction B [1]. To obtain these accuracies the

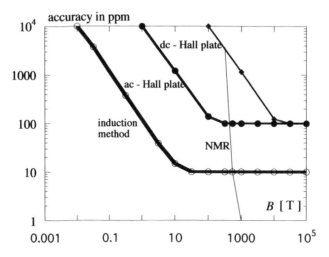

FIG. 13.1. Accuracies of different measuring methods.

use of modern digital integrators, voltage-to-frequency converters VFC, fast angular decoders, and of coils wound with high precision are essential.

13.2.1 The rotating or harmonic coil

Rotating or harmonic coils are conceived for direct measurements of two-dimensional field harmonics, which can be derived from a complex potential Φ [2]—see also eqns (9.52)–(9.54):

$$\begin{aligned}
\Phi &= V - i\,\vec{A}_z \\
B^* &= B_x(z) - iB_y(z) = B_\rho - iB_\theta
\end{aligned} \tag{13.1}$$

with

$$\frac{\delta\Phi(z)}{\delta z} = -\sum_{n=1}^{\infty} B_{\text{ref}}\, c_n \left(\frac{z}{R_{\text{ref}}}\right)^{n-1} \tag{13.2}$$

where

$$z = x + iy = r\exp(i\theta) = r(\cos\theta + i\sin\theta) \tag{13.3}$$

and

$$c_n = a_n + ib_n = \left(\frac{\Delta B_n}{B_{\text{ref}}}\right); \ |z| = R_{\text{ref}} \tag{13.4}$$

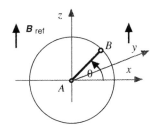

FIG. 13.2. 2D flux representation seen by a rotating coil in a dipole field.

stand for the normalized field harmonic errors related to B_{ref} at R_{erf}. Equation (13.2) yields:

$$-\Phi(z) = R_{ref} B_{ref} \sum_{n=1}^{\infty} \frac{1}{n} c_n \left(\frac{z}{R_{ref}} \right)^n.$$ (13.5)

The integrated voltage VS of a rectangular rotating **radial coil** with conductors at points A, B according to Fig.13.2, measured in the $z = x + iy$ plane is equal to [3]:

$$\frac{VS}{N_t L_{eff}} = \int_A^B B(z)dz = R_{ref} B_{ref} \sum_{n=1}^{\infty} \left(\frac{R_m}{R_{ref}} \right)^n \frac{1}{n} (b_n \cos n\theta + a_n \sin n\theta).$$ (13.6)

It is convenient to make the coil radius of measurement $R_m = R_{ref}$; the integrated voltage then amounts to:

$$vs(\theta) = \frac{VS(\theta)}{N_t L_{eff} B_{ref} R_{ref}} = \sum_{n=1}^{\infty} \frac{1}{n} (b_n \cos n\theta + a_n \sin n\theta)$$

$$= \operatorname{Im} \left[\sum_{n=1}^{\infty} \frac{c_n}{n} \exp(in\theta) \right]$$ (13.7)

where b_n and a_n are the normal and skew field multipole errors at $R_m = R_{ref}$, N_t is the number of turns, and L_{eff} is the effective coil length. The coil effective surface is:

$$S_{eff} = N_t L_{eff} (R_m - R_{int}).$$ (13.8)

It is convenient to place the rotating coil inner radius on the rotation axis: $R_{int} = 0$. The consequence of choosing $R_{ref} > R_m$ is shown in Fig. 13.3: the measuring error for higher field harmonics increases rapidly in the case of an extrapolation of the measurements beyond R_m.

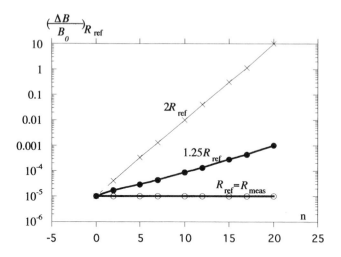

FIG. 13.3. Accuracy of measured field harmonics depending on the choice of R_{meas} and R_{ref}.

For a quadrupole magnet of field gradient G_0 [T m^{-1}], centred on the coil rotation axis one has:

$$B_x = G_0 y; \quad B_y = G_0 x \tag{13.9}$$

$$B_\theta = G_0 r \cos 2\theta; \quad B_r = G_0 r \sin 2\theta. \tag{13.10}$$

Integrating B_0 over the coil cross-section and introducing the geometrical factor

$$K_2 = N_t L_{eff}(R_{ext}^2 - R_{int}^2) \tag{13.11}$$

yields

$$vs(\theta) = K_2 \frac{G_0}{2} \cos 2\theta. \tag{13.12}$$

A displacement of the rotating axis of the measuring coil by x_0, y_0 is equivalent to the addition in all points of a dipole field according to eqns (13.9). and (13.10). The integrated flux is then:

$$vs(\theta) = K_2 \frac{G_0}{2} \cos 2\theta + S_{eff} G_0 r_0 \cos(\theta - \alpha) \tag{13.13}$$

$$r_0 = (x_0^2 + y_0^2)^{0.5}; \quad \alpha = \text{arctg} \frac{y_0}{x_0}. \tag{13.14}$$

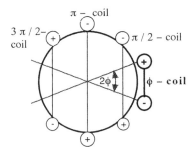

FIG. 13.4. Tangential coil assembly with dipole and quadrupole rejection coils.

In addition or as an alternative to radial coils, rotating **tangential coils** can also be used. An assembly of tangential coils located on the circumference of a rotating cylinder is shown in Fig. 13.4. The integrated flux seen by the 2φ coil is:

$$vs(\theta) = \mathrm{Im} \left\{ \sum_{n=1}^{\infty} \frac{c_n}{n}[\exp(in\theta + in\varphi) - \exp(-in\theta + in\varphi)] \right\}$$

$$= \sum_{n=1}^{\infty} \frac{2\sin n\varphi}{n}(a_n \cos n\theta - b_n \sin n\theta). \tag{13.15}$$

The coil geometrical factor is

$$k_n = 2\sin n\varphi. \tag{13.16}$$

In a tangential coil the polarity of the normal component b_n is inverted, compared to a radial coil. The main dipole and quadrupole terms can be rejected or compensated by a π coil and by two $\frac{\pi}{2}$ and $\frac{3\pi}{2}$ coils. The number of turns of the compensating coils are:

$$N_{t\pi} = N_{t\varphi} \sin \varphi \tag{13.17}$$

$$N_{t\frac{\pi}{2}} = N_{t\frac{3\pi}{2}} = \frac{\sin 2\varphi}{2} N_{t\varphi}. \tag{13.18}$$

The tangential coils, located on the cylinder surface, provide high rigidity to the frame against bending and torsion. The accurate positioning of the coils is more complicated and requires fine resistor trimming of the rejection ratio. Figure 13.5 shows a typical radial harmonic coil setup, consisting of several coils, 2–3 short ones to measure the central and the end field regions plus a long coil of $L_{\mathrm{eff}} > L_{\mathrm{magn}}$ to measure the integrated field and the equivalent length L_{m}. Figure 13.6 shows the scheme of a modern digital integrator with a VFC. Earlier harmonic coil systems did not allow continuous reading and the coils had to rotate stepwise, returning after each step to the initial zero position. The mechanical, electronic, and other components of a measuring system exhibit tolerances, which will result in measuring errors. Detailed analysis of the error

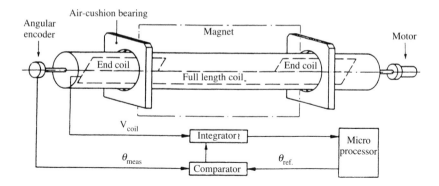

FIG. 13.5. Schematic diagrams of a harmonic coil measuring system.

sources and of their compensation have been made by L.Walckiers, P. Schmuser and W.G. Clark [3, 4, 5]. Modern continuously rotating measuring systems are nowadays using fast shaft (or angular) encoders with a time resolution better than $10\mu s$, low noise and low drift amplifiers of 10 nVs resolution, and VFCs operating at frequencies above 100 kHz. An amplitude resolution of better than 10^{-5} can be obtained. The obtained flux curve is analysed by a fast Fourier transform at 2^N points ranging from 124 to 1024. The errors of the measuring system are due to the quality of the shaft encoder, the torsion of the coil frame, the electronic system linearity, the quality of the main harmonic compensation, the displacement of the measuring system axis with respect to the magnet axis, the misalignment of the measuring coils, and the finite dimensions of the measuring coils. To obtain a higher resolution for the higher multipole terms, so-called bucking coils are used to compensate the dipole and quadrupole terms. Figure 13.7 shows the respective coil arrangements, while Fig. 13.8 shows a three-coil assembly to determine the misalignment between the coil axis and the axis of rotation. To conclude, harmonic coils are particularly suited for magnetic measurements of superconducting magnets for accelerators and colliders: during one full turn, taking a few seconds, precise data on the field quality, the main field orientation, the field axis and the amplitudes of the field harmonics are obtained.

13.2.2 *Field measurements with Hall probes*

The basis of magnetic field measurements with Hall-probes is the Hall effect, discovered in 1856 by W. T. Thomson and explained by E. H. Hall [6] in 1879. As shown in Fig. 13.9, the effect is due to the interaction of a magnetic field B perpendicular to a unidirectional current I or to the movement of charged particles; a Hall voltage u_H is then induced which is orthogonal to the B-I plane. The Hall voltage u_H or the electric field E_H are created by the current density $\vec{j_y}$ in the Hall plane, perpendicular to the current density $\vec{j_x}$ of the longitudinal control current I. u_H can be measured between

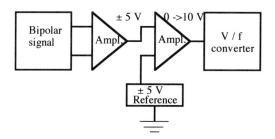

FIG. 13.6. Simplified block diagram of a modern digital integrator for magnetic measurements.

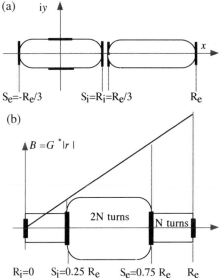

FIG. 13.7. (a) Bucking coils rejecting the main dipole field. Coils S (S_i, S_e) and R(R_i, R_e); (b) coil assembly rejecting both dipole and quadrupole harmonics of a quadrupole magnet.

two electrodes at a distance b as [7]

$$u_H = -bE_x\mu_H B = -I\frac{\mu_H}{d\,\sigma}B \tag{13.19}$$

with the Hall mobility coefficient μ_H [$m^2 V^{-1}s^{-1}$] given by

$$\mu_H = \frac{q t_{col}}{m} \tag{13.20}$$

where q and m are the charge and mass of the moving particle and t_{col} is the average time between two collisions. One can also define the Hall coefficient H [m^3 A^{-1} s^{-1}]:

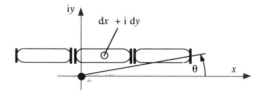

FIG. 13.8. Three-coil assembly for the determination of the misalignment between the coils and rotation axis.

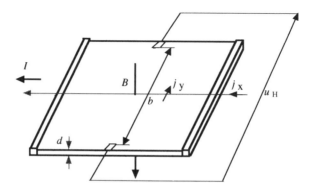

FIG. 13.9. Principle of a Hall generator or Hall plate.

$$H = \frac{\mu_H}{\sigma}. \tag{13.21}$$

An efficient Hall plate must be thin and exhibit high charge mobility at low electric conductivity σ, such as with the semiconducting compounds GaAs, InAs, or InSb.

Hall probes exhibit important thermal effects of a thermoelectric and thermomagnetic nature. It is therefore important to stabilize the Hall plate temperature to better than $T_{cal} \pm 0.1K$. In addition the magnetic field dependence of a Hall probe must be calibrated, preferably by high-precision NMR devices. Temperature stabilization is also important to reduce the temperature dependence of the Hall offset voltages in the range 0.2–$300\ \mu VK^{-1}$. Improved linearity in u_H is obtained by using cruciform shaped Hall probes with four-fold symmetry. The sensitivity can further be improved by ac current excitation. [1, 7].

In addition to temperature dependent and the non-linear calibration curves limiting the accuracy of Hall-probe measurements, the so-called planar Hall effect and the de Haas–Šubnikov effect, appearing at cryogenic temperatures, are important. The planar effect limits the measuring accuracy of low perpendicular fields B in the presence of a strong planar field, parallel to the plane of the Hall probe. Hall plates for low temperature measurements are nowadays commercially available; their measuring accuracy is ∼ 1%. At high fields increased field-dependent oscillations appear. This is the de Haas–Šubnikov effect. For field measurements in superconducting magnets it is thus advisable to place a room temperature anticryostat in the cold bore and perform Hall-probe measurements at ambient temperature. Well-designed and stabilized Hall probes in conjunction with fast, high-precision digital voltmeters can attain a long time accuracy of ∼ 100 ppm; a ∼ 30 ppm accuracy over several months can be obtained if the sensors are permanently powered.

To conclude, Hall probes are a valuable tool for magnetic measurements, provided all relevant error sources are taken into account and adequately compensated. Hall probes can then be used for point-to-point measurements in dc, ac, or pulsed multipole magnets and also for end field measurements. They are complementary to rotating coil measurements. Finally Hall probes are well suited for magnetic field measurements in large volume particle detector magnets.

REFERENCES

Bibliography to Chapter 2

[1]H. K. Onnes. Communications of the University of Leiden, **No 108**; *Proc. of the Royal Academy, Amsterdam*, **No 11, p.168**, 1908.

[2]H. K. Onnes. Communications of the University of Leiden, **No 108**; *Proc. of the Royal Academy, Amsterdam*, **No 1206**, 1911.

[3]W. Meissner and R. Ochsenfeld. *Naturwissenschaften* **No 21**, p. l206, 1933.

[4]C. J. Gorter and H. Casimir. *Phyxica* **I. p. 306**, 1934.

[5]F. London and H. London. *Zeitschrift fuer Physik*, **Vol.96, No 359**, 1935; and *Une conception nouvelle de la superconductivite* Paris, 1937, Edition Hermann & Cie.

[6]L. D. Landau. *Phys. Z. Soviet*. **No 11, p. 545**, 1937.

[7]L. D. Landau and E. M. Lifschitz. *Course of Theoretical Physics*, **Vol.5**, p.43, Pergamon Press 1959.

[8]J. Bardeen, L. N. Cooper, J. R. Schrieffer. *Phys. Review* **No 108, p.1175**, 1957.

[9]R. Doll and M. Näbaur. *Phys. Review Letters* **No 7, p.51**, 1961.

[10]B. S. Deaver and W. M. Fairbank. *Phys. Review Letters* **No 7, p. 43**, 1961.

[11]A. A. Abrikosov. *Zh. Exp. & Theor. Phys*; **No 32, p. 1442**, 1957; and *Sov. Phys*. JETP **No 9**.

[12]U.Essmann and H.Träuble. *Phys. Letters* **24a**, p.526, 1967; and *Scientific Instruments* **No 43, p. 344**, 1966.

[13]C. P. Bean. Magnetization of Hard Superconductors; *Phys. Review Letters*, **No 8, p. 250**, 1962.

[14]Y. B. Kim, C. F. Hempstead, and A. R. Strand. Magnetization and Critical Super currents; *Phys. Review* **No 129**, 1963.

[15]Andersen *Phys. Rev.* **No 124, Vol. 41**, 1961.

Bibliography to Chapter 3

[1]P. F. Chester. *British Patent Specification* **No 1**, 124, 622, 1964.

[2]P. F. Chester. *Proc. First Cryogenic Conference*, Tokio, **196a**, p.147–9, ed. K.Mendelsohn, Heywood-Temple publication.

[3]R. Hancox. *Proc. IEE* **Vol. 113**, p. 1221–1228, 1966.

[3a]J. D. Lewin, P. F. Smith, C. R. Walters, M. N. Wilson. Experimental and theoretical studies of filamentary superconducting composites; *Rutherford Magnet Group report RPP/73*, November 1969.

[4]R. Hancox. *Proc. IEEE Trans. on Magnetics*, **MAG-4** p. 486–488, 1968.

[4a]J. D. Lewin, P. F. Smith, C. R. Walters, M. N. Wilson. *Nuclear Instr. & Methods*, **Vol. 52**, p. 298–308, 1969.

[5]J. L. Duchateau, B. Turck. Prediction of maximum quenching current multifilamentary composites or multistrand cable; *Proc. of the Fifth Conf. on Magnet Techn., Frascati, Italy*, April I975.

[6]J. L. Duchateau, B. Turck. *Cryogenics* **No. 14, Vol. 9**, p. 481, 1974.

[7]C.P. Bean and P. S. Swartz. *Journal of Appiled Physics* **No 39, Vol. 11**, p. 4991, 1968.

[8]M. Scherer. Influence of the normal conducting matrix on the current repartition and current carrying capacity in technical superconductors (in German). *KFK-Report 2949*, Karlsruhe 1980.

[9]M. N. Wilson. *Superconducting Magnets.* Oxford University Press, Oxford, 1983.

[10]H. S. Carslaw and J. C. Jaeger. *Conduction of heat in solids.* Oxford University Press, Oxford, 1959.

Bibliography to Chapter 4

[1]P. W. Anderson. *Phys. Review Letters* **9**, p. 309, 1962.

[2]CERN LEP Design Report, CERN-LEP/**84-01**, June 1984.

[3]SSC-Study Group Site-Specific Conceptual Design of the Superconducting Super Collider, *SSC-Rep.* **SR-lO51**, June 1990.

[4]CERN-LHC Study Group The Large Hadron Collider; Conceptual Design, CERN/AC/**95-05**, (LHC) 20 October 1995.

[5]W. Buckel. Supraleitung; Grundlagen und Anwendung; *Physik Verlag GmbH, Weinheim, Germany* 1972.

[6]M. N. Wilson. Filamentary composite superconductors for pulsed magnets; *Applied Division, Rutherford Laboratory Report* RPP/A, March 1989.

[7]A. Gosh Magnetization measurements of Nb_3Sn multifilamentary wires; paper presented at the *High Field Accelerator Magnet Workshop, LBL Berkeley, USA*, 09–11 March 1993.

[8]G. H. Morgan. The Bean Model of A.C. Losses in Type II Superconductors; *BNL Report AADD-***146**, December 1968.

[9]W. Carr. *IEEE Trans. on Magnetics, MAG* **13 (1)**, p. l92, 1977.

[10]M. N. Wilson. Rate dependent magnetization in flat, twisted superconducting cables *RHEL Report M/A* **26**, Sept. 1972.

[11]L. Krempasky. A.C. Iosses in flat, twisted superconducting cables; *CERN-Report SPS/EA* **78-2**; February, 1978.

Bibliography to Chapter 5

[1]H. Hillmann and H. Krauth. NbTi based superconductors: Technical aspects and trends; *ICMC-Conference, Cambridge, UK* 1985, Paper **BZ-7**.

[2]D. C. Larbalestier. *et al.* High critical current densities in industrial scale composites

made from high homogeneity Nb 46.5 % Ti; *IEEE Trans. on Magnetics*, Vol. **Mag. - 21, No 2**, March 1985.

[3]S. Foner and B. Schwartz. Superconducting Material Science; Metallurgy, Fabrication & Applications. *Plenum Press, New York & London*, 1981–1982.

[4]P. J. Lee, C. McKinnell and D. C. Larbalestier. Restricted, novel heat treatments for obtaining high j_c in Nb 46.5 wt % Ti. ASC, University of Wisconsin, Madison; *publ. in Advances in Cryog. Engineering*, **Vol. 36**, Plenum Prexs, New York, 1990.

[5]Hoang Gia Ky. *et al.* Les conducteurs supraconducteurs a base de NbTi. *Reprint Alsthom - Atlantique, Belfort*, 1984.

[6]H. Krauth. Development and large scale production of NbTi and Nb_3Sn conductors for beam line and detector magnets; Rep **VI - A - 7**, 1990, *IISSC, Miami Beach, Florida*, March 1990.

[7]H. Krauth *et al.* An updated comparison of NbTi and Nb_3Sn supraconductors for fusion devices. *16th Symposium on Fusion Technology*; 3–7 September 1990, London.

[8]R. M. Scanlan *et al.* Evaluation of APC NbTi superconductor in a model dipole magnet; Report **SC - Mag 377, LBL - 32073**, August 1992.

[9]R. M. Scanlan *et al.* Characterization and coil test results for a multifilamentary NbTi conductor using artificial pinning center technology; Report **SU- Mag. 377, LBL - 32073**, 1992.

[10]P. A. Hudson *et al.* The critical current density of filamentary Nb_3Sn as a function of temperature and field. *IEEE Trans. on Magnetics*, **Mag - 19, No 3**, May 1983.

[11]J. E. Kunzler *et al. Physics Review Letters*, **Vol. 6, p. 89**, 1961.

[12]J. J. Hanak *et al. RCA Review* **No 25, Vol. 42**, 1964.

[13]E. J. Saur *et al. Proceedings of the International Conf. on High Magnetie Fields*; **p. 589**, 1961.

[14]M. G. Benz. *IEEE Trans. on Magnetics*, **MAG-2, p. 760**, 1966.

[15]H. Hillmann *et al.* Filamentary A15 Superconductors; Editors M. Suenaga & A. F. Clark, *Plenum Press, New York*, **p.17**, 1980.

[16]H. Krauth. Development of bronze route Nb_3Sn superconductors for accelerator magnets; *ICFA Workxhop on Superconducting Magnets and Cryogenics*, May 12–16, 1986.

[17]H. Krauth. Properties of NbTi and Nb_3Sn fine filament conductors; *Paper presented at the EPAC 90 (Second European Particle Conf.) Nice, 12–16 June 1990, Edition Frontieres.*

[18]M. Thoener *et al.* Nb_3Sn filamentary superconductors; an updated comparison of different manufacturing routes. *IEEE Trans on Magnetics*, **Mag-27, No 2**, March 1991.

[19]Seung Hong. High field conductor fabrication at Oxford Superconductor Techn. *Paper presented at the High Field Accelerator Magnet Workshop at LBL, Berkeley Ca*, 9–11 March 1993.

[20]Dingan Yu. High field conductor fabrication status at IGC, *Paper presented at the High Field Accelerator Magnet Workshop at LBL, Berkeley Ca*, 9–11 March 1993.

[21]E. Gregory. High fileld conductor fabrication status at IGC; *Paper presented at the High Field Accelerator Magnet Workshop at LBL, Berkeley Ca*, 9–11 March 1993.

[22]J. Mc Kinnell. High field conductor fabrication status at TWCA; *Paper presented at the High Field Accelerator Magnet Workshop at LBL, Berkeley Ca*, 9–11 March 1993.

[23]E. M. Hornsveld *et al. Adv. Cryogenic Engineering*; **Vol. 34 p. 493–498**, 1988.

[24]J. E. C. Williams *et al.* Conductors for a 1 GHz superconducting magnet; *IEEE Trans. on Magnetics*, **Vol. 30, No 4**, 1994.

[25]W. D. Markiewicz *et al.* 20 T model coil for a very high spectrometer magnet; *Paper presented at the CEC-ICMC*, 1991 Conf.

[26]G. Rupp. Improved evidence on the current behaviour in multifilamentary Nb_3Sn conductors under stress. *Trans. IEEE on Magnetics* **Vol. 13, No. 5**, September 1977.

[27]W. Specking *et al.* Effects of transverse compression on I_c in Nb_3Sn multifilament wires *Preprint ICMC of the Kernforsch. Zentrum Karlsruhe*, 1987.

[28]J. W. Ekin. Effects of transverse compressive stresses on the critical current of Nb_3Sn; *Paper prexented at the USA Dept of Energy Workshop on Nb_3Sn, Cambridge, Mass.* **August 1988**.

[29]B. Jakob & G. Pasztor. Effects of the transverse compressive stress on the critical current of cabled Nb_3Sn conductors. *Paper M-3, Applied Superconductivity Conference* **ASC - 1988**, San Francisco, August 1988.

[30]M. Bona *et al.* Reduced sensivity of Nb_3Sn-epoxy impregnated cables to transverse stress; *LHC Note* **141**, *CERN/AT MA/* **91-02**.

[31]H. Ten Kate. Strain-critical current measurements on Nb_3Sn cables; *Paper presented at the High Field Accelerator Magnet Workshop LBL, Berkeley, USA*, **9-11 March l993**.

[32]J. M. Van Oort *et al.* The reduction of the critical current under transverse pressure in a new cable TWCA-MJR cabled conductor. *Report SC - MAG -***414**, *LBL* **33430**, July 1993.

[33]P. Rabiller. Promising advances in the elaboration of superconducting Pb Mo6 S8 based wires; Materials Letters **15**, p. 19–25, 1992 North Holland Ed.

[34]O. Fischer. Superconducting, microstructural and grain boundary properties of hot-pressed Pb Mo_6 S_8; *Journal of Applied Physics*, **72(9)**, 1992.

[35]Evetts Concise Encyclopedia of Magnetic and Superconducting Materials. *Ed. J. Evetts*, 1992.

[36]M. Decroux *et al.* Overview on the recent progress of Chevrel phases and their impact on the development of Pb Mo_6 S_8 wires. *Proc. of the 1992 Applied Superconductivity Conference; Trans. IEEE*, **p. 1502**, 1993.

[37]Internal Workshop on Chevrel phase superconductors; Chavanne-de-Bogis, Switzerland.

[38]R. Fluckiger *et al.* Hot rolling of silver-sheathed Bi (2223) tapes. Physics **C 216, p. 339–344**; North Holland, 1993.

[39]R. Fluckiger *et al.* Pressed and cold-rolled, Ag sheathed Bi (2223) tapes; a comparison. *Physica C* **217, p. 335–341**; North Holland, 1993.

[40]R. Fluckiger. Hoch -T_c - Supraleiter in der Energietechnik; *Bull. SEV/VSE* **No 11**, 1994.

[41]D. C. Larbalestier and B. Riley. Prospects for the use of high temperature superconductors in accelerators; *Eloisatron Project Workshop* **29**; *Superconducting Accelerator and Defector Magnets at extremely high fields*, Erice, Italy, May 1993.

[42]T. Shintomi. Are HTC superconductors available for ultra high field accelerator magnets? *Eloisatron Project Workshop* **29**; *Superconducting Accelerator and Defector Magnets at extremely high fields*, Erice, Italy, May 1993.

Bibliography to Chapter 6

[1]G. Horlitz *et al.* Report on the Performance and Reliability of the Hera Cryogenic System. *Advances in Cryogenic Eng.* **Vol. 39**, 1994.

[2]SSC Laboratory Site-Specific Conceptual Design of the Superconducting Supercollider; *Report SSC-SR* **1051**, June 1990.

[3]H. H. Holm. A closed loop system for superconducting bubble chamber magnets; *Proc. of the Int. Symposium on Magnet Teehnology*, **p. 611**, Stanford, 1965.

[4]M. Morpurgo. Construction of a superconducting test coil, cooled by forced helium circulation. *CERN Report* **68-17**.

[5]G. Ries. Theoretical studies on stability in circulation. *CERN Report* **68-17**.

[6]G. Ries. Theoretical studies on stability in forced flow cooled superconductors 'Stability of Superconductors' International Institute of Refrigeration; Commission A/2*, **1981–6**. Saclay, France.

[7]US National Bureau of Standards Thermophysical Properties of Helium 4 from 2 to 1500 K with pressures to 1000 bar; *Technical Note* **631**, November 1972.

[8]H. Brechna. Properties of solid materials affecting the stability of current carrying devices; *Stability of Superconductors, Int. Institute of Refrigeration, Comm. A 1/2, Saclay*, **1981–6**.

[9]C. Schmidt. Review of steady state and transient heat transfer in pool boiling helium I; *Stability of Superconductors, Int. Institute of Refrigeration, Comm. A 1/2, Saclay*, **1981–6**.

[10]C. Johannes. Studies of forced flow convection heat transfer to helium I. *Cryogenic Engineering Conference Boulder, Colorado*, **17–19 June 70**.

[11]Z. Stekly *et al.* Stable Superconducting Coils; *IEEE Transactions on Nuclear Science*, **Vol. 12, page 367**, 1965.

[12]B. J. Maddock *et al.* Superconductive Composites—Heat transfer and steady state stabilization; *Cryogenics 1969*, **Vol. 9**.

[13]Ch. Meuris. Influence of an uncooled region on the stability of superconductors.

Stability of Supereonductors; Int. Institute for Refrigeration, Comm. A 1/2 Saclay, France **1981–6**.

[14]Bejan, A. and Tien, C. L. *Cryogenics*, **Vol. 8, p. 433**, 1978.

[15]Turowski, P. A review on stability experiments with superconductors in liquid He under pool boiling conditions. *Cryogenics*, **Vol. 8, p. 433**, 1978.

[16]Baynam, D. E. *et al. IEEE Trans. on Magnetics*, **No. 17, p. 732**, 1981.

[17]Pasztor G. and Schmidt C. *Journal of Applied Physics*, **Vol. 49, p. 886**, 1978.

[18]Nick, W. Theoretical studies of stability in pool boiling helium I; *Stability of Superconductors, Int. Institute for Refrigeration, Comm. A1/2* Saclay, France, **1981–6**.

[19]Greene, W. and Saibel, E. Stability of internally cooled superconductors; *Advances in Cryogenic Engineering*, **Vol. 14, p. 138**, 1974.

[20]Hoenig, M. O. and Montgomery, D. B. Dense supercritical helium cooled superconductor for large high field stabilized magnets. *IEEE Trans. Magn.* **Vol. 11, p. 569**, 1974.

[21]Dresner, L. *et al. Proc. of the eighth Int. Cryogenic. Conf; IPC Science & Technology Press Ltd; Guildford, England*, **p. 32, 1980**.

[22]Krafft, G. *et al. Proc. of the eighth Sympoxium on Engineering Problems in Fusion Research, IEEE Science Center Piscataway, N.J.* **Vol. IV, p. 1724**, 1979.

[23]Claudet, G. *et al.* The design and operation of a refrigerator system using super fluid helium; *Proc. Fifth Int. Cryogenic Engineering Conf.* **p. 265–267**, Kyoto, 1974.

[24]Claudet, G. and Seyfert, P. Bath cooling with subcooled superfluid helium; *paper presented at the Cryogenic Engineering Conf.* **1981**, San Diego, California.

[25]Association Euratom-CEA Tore II Supra; *Report EUR. - CEA* **Fc 1021**, 1979.

[26]Khalatnikov, I. M. In Introduction to the theory of superfluidity; *ed. W. J. Benjamin*, New York, 1966.

[27]Vinen, M. F. Physical properties of superfluid helium; *Stability of Superconductors, Int. Institute for Refrigeration, Comm. A 1/2*, **1981–6**, Saclay, France.

[28]Van Sciver *et al.* Heat transfer to He II in cylindrical geometries; *Advances in Cryogenic Engineering*, **Vol. 25, p. 363–371**, 1980.

[29]Van Sciver, S. W. Heat transfer in superfluid He II; *Proc. of the Eighth Int. Cryogenic Conf., IPC Press, Genova*, **p. 228**, 1980.

[30]Seyfert, P. Practical results on heat transfer to superfluid helium; *Stability of Superconductors, Int. Institute for Refrigeration, Comm. A 1/2*, **1981–6**, Saclay, France.

[31]Frederking, T. H. K. Transient superconductor (NbTi Cu) - He II heat transfer rates to pressurized superfluid during step input in power; *Stability of Superconductors; Workshop; Int. Inst. of Refrigeration, Comm. A 1/2* **1981–6**, Saclay, France.

[32]Meuris, Ch. Transient stability of superconductors, cooled by superfluid helium at atmospheric pressure; *Stability of Superconductors; Workshop; Int. Inst. of Refrigeration, Comm. A 1/2* **1981–6**, Saclay, France.

[33]Van Sciver, S. W. Transient heat transfer in helium II; *Cryogenics*, **Vol. 19, p. 385**, 1979.

[34]Seyfert, P. Transient heat transfer into boiling He I and unsaturated He II. *Adv. in Cryogenic Engineering*, **Vol. 25, p. 378, 1980**.

[35]The LHC Study Group of CERN The Large Hadron Collider; Conceptual Design; *Report CERN/AC / 95-05* **(LHC)**, 20 Oct. 1995.

[36]Meuris, Ch. and Leroy, D. Transfert thermique dans l'isolation de cables supra-conducteurs d'accèlèrateurs refroidis par helium superfluide. *4 - èmes Journées d'Aussois*, **1993**.

[37]Meuris, Ch. Heat transport in the insulation of cables, cooled by superfluid helium. *Symposium on Superconductor Stability*, Yokohama Japan, **Nov 1990**.

[38]Roubeau, G. Principle of the Roubeau bath, *CF C R Acad. Sc. Paris*, **273, B 581–583**, 1971.

[39]S. W. Van Sciver. Helium Cryogenics. *The Int. Cryogenics Monograph series*; **1986 Plenum Press**, New York.

Bibliography to Chapter 7

[1]Wilson, M. N. The Quench Computer Program *Rutherford Laboratory Report* **RHEL/151**, 1968.

[2]Koepke, K. TMAX Computer Program of *Fermilab, Batavia, ll. USA*, 1975.

[3]Analog Inc. Oregon, USA SABER; *Saber Users Guide* **Analogy 1990**, Beaverton, Oreg.

[4]Rodriguez-Mateos, F. and Hagedorn, D. Quench calculation with QUABER *CERN report AT/ MA / FRM*. **Int. Note 90-14**.

[5]Hagedorn, D. and Rodriguez Mateos, F. Modelling of the quenching process in complex superconducting magnet systems. Paper presented at the *Magnet Technology Conference* **MT - 12**, July 1992, Lenjingrad, USSR.

[6]Wilson, M.N. Superconducting Magnets *Oxford University Press*, 1983.

[7]McInturff, M. D. Calculated quench conditions in TAP using QUCERN; *CERN Report SPS - EMA / 89 - 12*, June 1989.

[8]Mc Inturff, M. D. QUCERN Users' Manual; *CERN - SPS - EMA / 89-12*.

[9]Hagedorn, D. Thermal propagation of the normal region in s.c. windings; *CERN SPS - EA Note 76 - 28*, 1976.

[10]Rodriguez-Mateos, F., Szeless, B. and Calvone, F. Development of industrially produced quench heaters for the LHC superconducting lattice magnets. *LHC Project Report* **48, 23 September** 1996.

[11]Hagedorn, D. and Nagele, W. Quench protecting diodes for the LHC at CERN; Paper presented at the **1991** *CEC Conference* at Huntsville, Allabama.

[12]Rodriguez-Mateos, F. *et al.* Quench protection test results and comparative simulations on the first 10 m prototype dipoles for the Large Hadron Collider; *CERN /AT/95-22* **(MA)** *LHC Note* **333**.

[13]Coull, L. *et al.* LHC Magnet Quench Protection System, *CERN* **AT/ 93 - 42 (MA)**.

Bibliography to Chapter 8

[1]Schmuser, P. Basic Course on Accelerator Optics, *DESY - HERA Report* **87 - 02**. January 1987.

[2]Steffen, K. Basic Course on Acelerator Optics; CERN Accelerator School, Gyf - sur - Yvette, France. *CERN Yellow Report* **77-07**; November 1985.

[3]Wilson, E. J. N. Proton Synchrotron Accelerator Theory; *CERN Yellow Report* **77 - 07**; March 1977.

[4]Wilson, E. J. N. Circular Accelerators Transverse; *Report CERN /PS* **86-2**.

[5]Courant, E. D. and Snyder, H.S. Theory of the Alternating Gradient Synchrotron; *Anals of Physics*, **3, Vol. 1**, 1958.

[6]Christofilos, N. Focusing System for ions and electrons; *US Patent* **2, 736, 799** 1950. See also *CERN Yellow Report* 1995 - 06.

[7]Wilson, E. J. N. Non Linearities and Resonances. *CERN Yellow Report* **85-19**; November 1985.

[8]Wilson, E. J. N. Transverse Beam Dynamics; *CERN Yellow Report* **85-19**; November 1985.

Bibliography to Chapter 9

[1]The SSC Central Design Group Conceptional Design of the Superconducting Supercollider; *Lawrence Berkeley Lab. Report CDG-SSR-SR-* **2020**; March 1986.

[2]Wilson, E.J.N. Non linearities and resonances; *CERN Accelerator School, General Accelerator Physics; CERN Yellow Report* **85 - 19**; November 1985.

[3]Wolff, S. Superconducting Accelerator Magnet Design; *CERN Acc. School CAS; fifth general accelerator physics course, CERN Yellow Report* **94 - 01**; 26 January 94.

[4]Meuser, R. Symmetrical Intersecting Ellipse Quadrupole Magnets; *LBL - En gineering Note* **UCID - 3193**; July 1968.

[5]Asner, A. Cylindrical Aperture Multipoles with Constant Current Density Sector Windings; **CERN SI / MAE Note 69-15**; October 1969.

[6]Asner, A. Cylindrical aperture dipole and quadrupole fields obtained with sector windings of linearily decreasing current density; *CERN SPS - EMA - Note* **78 - 4**; February 1974.

[7]Russenschuck, S. ROXIE (the routine for the optimization of magnet crosssections, inverse problem solving and end region design;) *CERN AT- MA* **93 - 17**, *LHC- Note* **No 288**.

[8]Russenschuck, S. and Tortschanoff, T. Optimization of the coils of the LHC Main Dipole; *CERN AT-MA* **92 - 28**, *LHC Note* **206**.

[9]Meuser, R. Elimination of end effects in multipole magnets; *LBL Engineering Note* **UCID - 3510** April 1971.

[10]Iispeert, A. *et al.* A quick and exact method for the calculation of the field integrals over the coil ends; *CERN AT-MA* **91 - 07**; *LHC - Note* **153**, 1991.

[11]Asner, A. Magnetization currents in superconducting dipole and quadrupole magnet windings and their influence on the field precision; *CERN / I EA - Note* **74 - 2**, February 1974.

[12]Wolff, R. Persistent currents in LHC magnets; *IEEE. Trans. on Magnetics* **Mag. 28, No 1**, January 1992.

[13]Bruck, H. *et al.* Time dependence of persistent current effects in the superconducting Hera magnets; *Proceedings of the second European Particle Accelerator Conference, Nice,* **Vol. 1, p. 329**, June 1990.

[14]Bruck, H. *et al.* Observaton of a periodic pattern in the persistent current fields of the Hera dipole magnets; *DESY HERA report* **91 - 01**, 1991.

[15]Halbach, K. Fields and first order perturbation effects in two dimensional conductor dominated magnets; *Nuclear Instruments & Methods* **No 78, p. 185–198**, 1970.

[16]Orrell, D. The PE - 2D inverse magnetic field optimization program. *Private communication.*

[17]Focus Report IEEE Spectrum Software. *Publication of the Institute of El. and Electronics Engineers, Inc.* **No. 11, Vol. 28, November 1991**.

Bibliography to Chapter 10

[1]Leroy D. Private communication; 1994.

[2]Hagedorn, D. *et al.* Towards the development of high field superconducting magnets for a hadron collider in the LEP tunnel; *CERN LHC Note* **31/1985**.

[3]Asner, A. Vereinfachte mechanische Berechnung der Struktur des 250 Tm^{-2}, 6 cm-Bohrung, Nb_3Sn 'low β' Quadrupols; *DESY Technischer Bericht*, **October 1991**.

[4]Huette, II Band Mechanik, 29 *Ed. W. Ernst & Sohn*, **Berlin, 1971**.

[5]Timoshenko, S. Strength of Materials; Part II; Advanced theory and Problems; *Van Nostrand Reinhold Co*; **New York, Ed. August 1976**.

[6]Wiik, B.H. Design and Status of the Hera Superconducting Magnets; *DESY Hera Report* **88 - 05**; April 1988.

[7]Meinke, R. Superconducting Magnet System for Hera; *DESY - Hera Report* **90 - 17**; September 1990.

[8]Superconducting Super Collider Design Group Site Specific Conceptual Design; *Report*, **June 1990**.

[9]Dell' Orco, D. *et al.* A 50 mm bore superconducting dipole with a unique structure; *LBL Report* **1992**.

[10]Dell' Orco, D. *et al.* Design of the Nb_3Sn dipole D-20; *LBL report* **32072, SU - MAG 376**, 1992.

[11]Swanson Analysis Systems; ANSYS Computer Program Manual, **1993**.

[12]Verpeaux, Milliard, Charras, CEA Paris CASTEM2000; *Version* **1996**.

[13]Asner, A. *et al.* First Nb_3S 1 m long superconducting magnets for LHC break the 10 T field threshold; *Paper presented at the 11th Magnet Technology Conference*, 28 August–1 September, 1989. Tsukuba, Japan.

[14]Ten Kate H., den Ouden, A. The Nb₃Sn dipole project at the University of Twente, *Paper presented at the workshop of high field magnets*; **Erice, Italy, 8–18 March, 1995**.

[15]Bacconier, Y. *et al.* LHC; The Large Hadron Collider Accelerator Program At CERN; *CERN/AC/* **93-03**; November 1993.

[16]Russenschuck, S. A computer program for the design of superconducting accelerator magnets; *Paper presented at the ACES - Conference*, **March 1995**.

[17]Asner, A. *et al.* Towards a 1 m long, high field Nb₃Sn dipole of the ELIN–CERN collaboration; *IEEE Trans. On Magnetics*, **Vol. 25, p. 1636–1639**, March 1989.

[18]Asner, A. Die' Wind & React 'Technologie fuer Nb₃Sn - bewickelte Hochfeld-magnete; *Techn. Studie, DESY - Hamburg*, **July 1989**.

[19]Conte, R.R. Elèments de Cryogènie; *Edition Masson & Cie*, **Paris 1970**.

[20]Taylor, C., Scanlan R. Reports and Conclusions on the High Field Accelerator Magnet Workshop; *LBL, California, USA* **9–11 March 1993**.

[21]Scanlan R. Workshop on High Field Magnets for Accelerators; *LBL, Berkeley, USA*, **11–13 February, 1997**.

[22]Barletta, W. and Scanlan R. 29th Workshop on Superconducting Accelerator and Detector Magnets at Extremely High Fields; *Ettore Majorana Center, Erice, Italy*, **11–18th May, 1995**.

[23]McIntyre, P. The Pipe Magnet; High Field Accelerator Magnet Workshop; *Claremont Hotel, Berkeley, USA* **3–11 September, 1993**.

[24]Van Oort, J. M. and Scanlan, R. The pipe-quadropole, an alternative for high gradient interaction region quadrupole designs. **LBNL report 38331 SC MAG. 543**, August 1996.

[25]McIntyre, P. 16 Tesla Dual Dipole Development at TAMU. *Paper presented at the Workshop on High Field Accelerator Magnets*; LBNL, Berkley, USA, 11–13 February 1997

Bibliography to Chapter 11

[1]Horlitz, G. *et al.* Cryogenic test and operation of the superconducting magnet system in the HERA proton storage ring. *Advances in Cryogenic Engineering*, **Vol. 37 A, p. 653**, Plenum Press New York, 1992.

[2]Berg, H. *et al.* Report on operational experience and reliability of the HERA cryogenic system. *Advances in Cryogenic Engineering*, **Vol. 39**, 1994.

[3]SSC Laboratory Report Site specific conceptional design of the SSC. *SSC - SR - Report* **1051**, June 1990.

[4]Claudet, G. and Aymar, R. Tore Supra and the cooling of large high field magnets *Publ. in Advances of Cryogenic Engineering*, **Vol. 35 A, p. 55–67**, 1990.

Bibliography to Chapter 12

[1]Kunzler, J. E. *et al. Physics Review Letters*, **Vol. 6, p.68**, 1961.

[2]Durand, E. Electrostatique et Magnetostatique, *Edition Maason & Cie*, **Paris 1953**.

[3]Montgomery, B. D. Solenoid Magnet Design *Wiley - Interscience* **1969**.

[4]Wilson, M. N. Superconducting Magnets. *Oxford Univerxity Press*, **Oxford, 1983**.

[5]Chari, M. and Silvester, P. Finite elements in electromagnetic field problems. *Wiley & Sons*, **USA 1980**.

[6]Timoshenko, S. Strength of Materials Vols I and II*Van Nostrand Reinhold;* **Third Edition, 1955, N.Y.**

[7]Bloch, E. *Phys. Review*, **Vol. 70**; 1946.

[8]Markiewicz, D. High field magnet work at the National High Magnetic Field Laboratory; paper presented at the *High Field Accelerator Workshop* **9–11 March** 1993 at Berkeley, USA.

[9]Markiewicz, D. The 900 MHz/25 T NMR Program of the NHMFL, Florida, Tallahassee, USA *Report* **32306 4005**.

[10]Markiewicz, D. 20 T model coil for a very high field spectrometer magnet; *Paper presented at the* **1991 CEC / ICMC Conference**.

[11]Green M. *et al.* The LBL Time Projection Chamber - TPC *Advances in Cryogenic Eng.*; **Vol. 25, D - 4, p.194–199**, 1980.

[12]Desportes, H. *et al.* The CELLO and DESY detector solenoids; *Advances in Cryogenic Eng.*; **Vol. 25 D - 2, p. 175 184**; 1980.

[13]Andrews, D. *et al.* The Clio detector solenoid; *Advances in Cryogenic Eng.* **Vol 27 p. 143–150**; 1980.

[14]Minemura, H. *et al.* The Central Detector Facility - CDF-solenoid. *Nuclear Instrum. & Meth.* **A 238, p. 18–34**; 1985.

[15]Yamamoto, A. *et al.* The Topaz detector solenoid for KEK; *Proc. of the ninth Inter. Conf. on Magnet Technology*; **p. 167**; 1985.

[16]Tsuchiya, K. *et al.* The AMI - detector solenoid for KEK; *IEEE Trans. on Masgnets* **Mag. 23, No 2, p. 520–523**; 1987.

[17]Apzey, R. Q. *et al.* The Delphi detector solenoid for LEP - CERN; *IEEE Trans. Magn.* **Mag. 21, No 2, p.490–493**; 1985.

[18]Desportes, H. *et al.* Design, construction and test of the large superconducting solenoid Aleph; *IEEE Trans. on Magn.* **Vol. 24, No 2**, March, 1988.

[19]Desportes, H. *et al.* Conceptual design of the CMS 4 T - solenoid; *Report CENS, Saclay, Cryomag* **92 - 02 JCL GY February 1992**.

[20]Desportes, H. *et al.* Advanced features of very large superconducting magnets for SSC and LHC detectors; *Report CENS, Saclay, Cryomag* **83-10 HD / CV, MT-13 Conference** Victoria, Canada, 20–24 September 1993.

[21]Desportes, H. *et al.* 'Atlas' barrel superconducting toroid conceptual design; **CEA Report DAPHNIA / STCM 92 04**; October 1992.

[22]CERN LHC News **No 6**, December 1994

Bibliography to Chapter 13

[1]Henrichsen, K. N. Classification of magnetic measurement methods; *CERN Accelerator School on Magnetic Measurements and Alignment*; **16–20 March 1992, p. 71- 83**, Montreux, Switzerland.

[2]Briant, P. J. Basic theory for magnetic measurements, *CERN Accelerator School on Magnetic Measurements and Alignment*; **16–20 March 1992, p. 52–70**.

[3]Walckiers, L. The harmonic coil method; *CERN Accelerator School on Magnetic Measurements and Alignment*; **16–20 March 1992 p. 138–166**.

[4]Schmuser, P. Magnetic measurements of superconducting magnets and analysis of systematic errors; *CERN Accelerator School on Magnetic Measurements and Alignment*; **16–20 March 1992, p. 240–273**.

[5]Clark, W. G. Introduction to magnetic resonance and its application to dipole testing; *CERN Accelerator School on Magnetic Measurements and Alignment*; **16–20 March 1992**, p. 193–205.

[6]Hall, E. W. On a new action of the magnetic and electric currents; *Amer. Journal of Mathematics*, **2, p. 287–292**, 1879.

[7]Berkes, B. Hall generators; *CERN Accelerator School on Magnetic Measurements and Alignment*; **p.167–191; 16–20 March 1992**, Montreux, Switzerland.

INDEX